"互联网+"创新创业实践系列教材

软件项目开发实战 App全栈

钟元生　李普聪　主编
赵圣鲁　钟　坚　吴　凯　邹宇杰　秦　振　副主编

清华大学出版社
北京

内 容 简 介

本书围绕一个真实项目展开,通过一个个小案例,引导读者在较短时间内熟悉一个较大规模的App应用系统的开发,以培养App程序员的独立开发能力。

本书包括App项目需求分析、App应用体验、Android客户端设计和实现、数据库开发、Java Web服务器端设计、App客户端与服务器交互设计、Spring Boot服务器端设计和微信分享的实现等内容。

本书适合作为"App开发"实训课程教材,可供项目经验少的学生开发实用App时参考,也可作为大学生"互联网+"创新创业竞赛的实战指导书或软件类专业大学生完成"移动应用类毕业设计"的参考书。

本书封面贴有清华大学出版社防伪标签,无标签者不得销售。
版权所有,侵权必究。举报: 010-62782989, beiqinquan@tup.tsinghua.edu.cn。

图书在版编目(CIP)数据

软件项目开发实战: App全栈/钟元生,李普聪主编. —北京: 清华大学出版社,2022.9
"互联网+"创新创业实践系列教材
ISBN 978-7-302-61367-1

Ⅰ. ①软… Ⅱ. ①钟… ②李… Ⅲ. ①移动终端－应用程序－程序设计－高等学校－教材 Ⅳ. ①TN929.53

中国版本图书馆 CIP 数据核字(2022)第 124645 号

责任编辑: 袁勤勇
封面设计: 杨玉兰
责任校对: 李建庄
责任印制: 曹婉颖

出版发行: 清华大学出版社
网　　址: http://www.tup.com.cn, http://www.wqbook.com
地　　址: 北京清华大学学研大厦A座　　邮　编: 100084
社 总 机: 010-83470000　　邮　购: 010-62786544
投稿与读者服务: 010-62776969, c-service@tup.tsinghua.edu.cn
质量反馈: 010-62772015, zhiliang@tup.tsinghua.edu.cn
课件下载: http://www.tup.com.cn, 010-83470236

印 装 者: 三河市铭诚印务有限公司
经　　销: 全国新华书店
开　　本: 185mm×260mm　　印　张: 20.75　　字　数: 482千字
版　　次: 2022年9月第1版　　印　次: 2022年9月第1次印刷
定　　价: 59.00元

产品编号: 093711-01

前言

在"互联网+"创业公司中,App是标配。现有教材多数重点讲述服务器端、客户端以及数据库等其中某个部分,对应于软件项目组中的服务器程序员、Android/iOS程序员、数据库程序员等岗位,比较适合有一定规模的公司。目前许多培训机构以及高校都是按某一类岗位的要求来培养程序员,要找到这类程序员更容易。

很多时候,企业希望程序员同时承担设计服务器端、客户端和数据库等多个岗位的任务,即有全栈工程师的能力,初创公司尤其如此。即使在程序员分工较细的公司,如果程序员具有全栈工程师的基本能力,则不同岗位的程序员更易合作。

随着全国大学生"互联网+"创新创业竞赛的不断推进,许多学生都想将自己的好点子用App实现,但往往就卡在最终App的实现上,创意落地难。

对于教材来说,为满足全栈工程师的培养要求,则既要有一个完整的综合案例,又要能将案例拆分为一个个可独立运行、可测试的小例子。读者可以边学边做,检验自己的知识掌握程度,学完全书即完成完整项目。这类教材比较难设计,市面上少见,而本书正是想做这一尝试。

本书案例选自早期"豹考通"App的简化版,介绍了该App的Android客户端程序、服务器端数据库和服务器管理程序的开发。本书假定读者有一定的编程语言基础,对App开发和Android编程有一定的了解。

本书中有些代码较长,有些模块只列出关键代码。基于这些关键代码讲解,读者可以访问教学资料的网站,下载完整代码,边阅读代码边体验程序效果,并且扩充或修改程序功能。为方便阅读,书中的每段代码都添加了代码编号,部分关键语句加了注释,并给出了程序在资源包中的位置,样例如下所示。

程序清单 x-yy: Code040201\app\src\main\java\cn\jxufe\iet\code040201\ControLineActivity.java

```
1    private class MyItemSelectedListener implements AdapterView.OnItemSelectedListener {
2        @Override
3        public void onItemSelected(AdapterView<?> parent, View view,
4                                   int position, long id) {
5            switch (parent.getId()) {
6                case R.id.areaSpinner:        //如果是选择省份列表
7                    sourceAreaId = position;
8                    break;
9                case R.id.yearSpinner:        //如果是选择年份列表
                     ...
20           }
```

```
21              /*设置标题显示*/
22              controlLineTitle.setText(Html.fromHtml("<font color=red><b>"
                                                                    ...
```

其中1,2,3,…,22为代码行号,中间为代码实际内容,"//"后为注释内容。程序清单的位置也进行了详细标注,便于读者查找下载。各章代码独立存储在一个文件夹中。

阅读本书时,最好按照书中的步骤同时实际操作,加深印象,掌握关键技术,不建议仅直接打开源代码运行来查看最后效果。编程基础较弱者也可泛读全书,体验App开发流程和关键细节。涉及相关技术时,尽可能去参考相关教材或网络资源。

为方便学习、交流与资源共享,我们提供了本书相关资源的下载地址,网址为http://www.xs360.cn/book。

本书由钟元生、李普聪担任主编,负责全书的方案设计、内容策划、案例分解、细节把握、质量控制和统编定稿工作。各章分工如下:钟元生负责第1章、第2章,并参编其余各章,赵圣鲁负责第3章、第7章、第9章,参编第4章,邹宇杰负责第4章,吴凯负责第5章,钟坚负责第6章,秦振负责第8章。李浩轩参加了书稿校对、代码检查等工作。

作为一种尝试,我们在本书编写过程中反复对案例进行选择,对教材结构、教法、编法等进行研究与设计,历经一年半撰写了几十个版本,进行了三次大规模修改,本书方得以展现在读者面前。尽管如此,本书依然有许多不尽如人意的地方。还望读者批评指正,以便将来再版时完善。

本书适合作为本专科"App开发"课程实训教材,供无项目经验的学生开发App时参考,也可作为大学生"互联网+"创新创业竞赛的实战参考书。

编　者
于江西财经大学麦庐园
2022年6月

目 录

第 1 章　App 全栈开发概述　　<<<1
  1.1　什么是 App ………………………………………………………………… 1
  1.2　App 开发流程 ……………………………………………………………… 1
  1.3　案例介绍 …………………………………………………………………… 6
  1.4　本章小结 …………………………………………………………………… 7
  1.5　课后练习 …………………………………………………………………… 7

第 2 章　项目需求分析　　<<<8
  2.1　本章简介 …………………………………………………………………… 8
  2.2　功能需求分析 ……………………………………………………………… 8
    2.2.1　需求获取 ……………………………………………………………… 8
    2.2.2　功能分析 ……………………………………………………………… 8
    2.2.3　系统 UML 建模 ……………………………………………………… 10
  2.3　系统用例分析 ……………………………………………………………… 10
    2.3.1　系统用例图 …………………………………………………………… 10
    2.3.2　系统用例描述 ………………………………………………………… 11
  2.4　项目数据库分析 …………………………………………………………… 15
    2.4.1　数据库基本概念 ……………………………………………………… 15
    2.4.2　数据库设计的基本过程 ……………………………………………… 16
    2.4.3　系统涉及的实体及其属性 …………………………………………… 16
  2.5　本章小结 …………………………………………………………………… 20
  2.6　课后练习 …………………………………………………………………… 20

第 3 章　App 应用体验　　<<<21
  3.1　本章简介 …………………………………………………………………… 21
  3.2　开发环境配置 ……………………………………………………………… 22
    3.2.1　Java 语言 ……………………………………………………………… 22
    3.2.2　Java 环境配置 ………………………………………………………… 22
    3.2.3　Android Studio 下载 ………………………………………………… 29
    3.2.4　Android Studio 安装 ………………………………………………… 31
    3.2.5　创建第一个项目：HelloWorld ……………………………………… 35
    3.2.6　创建一个虚拟机设备 ………………………………………………… 36

3.3 本地数据版 App 案例 ·· 42
3.4 网络数据版 App 案例 ·· 45
 3.4.1 设计网络服务器 ·· 45
 3.4.2 Android 功能实现 ·· 47
 3.4.3 客户端运行效果 ·· 47
3.5 本章小结 ·· 49
3.6 课后练习 ·· 49

第 4 章 Android 客户端设计和实现   <<< 51
4.1 本章简介 ·· 51
4.2 "查询界面"模块设计与实现 ·· 51
 4.2.1 用 Spinner 实现下拉列表选项 ·· 52
 4.2.2 事件监听器 ·· 55
 4.2.3 ListView 列表 ·· 56
4.3 "报考咨询"模块设计与实现 ·· 62
 4.3.1 界面设计 ·· 62
 4.3.2 问题列表的实现 ·· 66
 4.3.3 问题回复对话框的实现 ·· 71
4.4 "个人基本信息"模块设计和实现 ·· 73
 4.4.1 界面设计 ·· 73
 4.4.2 用 SharedPreferences 实现个人信息存储 ·· 76
 4.4.3 个人基本信息填写对话框的实现 ·· 78
4.5 多页面切换效果设计与实现 ·· 86
 4.5.1 多页面切换框架的实现 ·· 87
 4.5.2 为选项卡添加对应内容 ·· 94
 4.5.3 多页面内容填充 ·· 100
4.6 绘制趋势图 ·· 108
 4.6.1 绘制图类 ·· 109
 4.6.2 用绘图类绘制坐标轴 ·· 111
 4.6.3 用绘图类绘制趋势线——源数据来自数组 ·· 118
 4.6.4 用绘图类绘制趋势线 ·· 131
 4.6.5 编写趋势线相关 Activity 和 Fragment ·· 135
4.7 本章小结 ·· 142
4.8 课后练习 ·· 142

第 5 章 数据库开发   <<< 144
5.1 本章简介 ·· 144
5.2 SQLite ·· 144

5.2.1　了解 SQLite ·· 144
　　　5.2.2　安装 SQLite ·· 145
　5.3　创建 SQLite 数据库 ·· 148
　　　5.3.1　创建 bkt 数据库 ··· 148
　　　5.3.2　创建 area 表 ··· 148
　　　5.3.3　插入 area 表数据 ·· 150
　　　5.3.4　创建 school 表 ··· 151
　　　5.3.5　导入 school 表数据 ·· 151
　5.4　SQLite 数据库操作类与接口 ·· 157
　　　5.4.1　SQLiteDataBase 类 ·· 157
　　　5.4.2　SQLiteOpenHelper 类 ·· 158
　　　5.4.3　Cursor 接口 ·· 158
　　　5.4.4　ContentValues 类 ··· 158
　5.5　从 SQLite 获取 ListView 列表项的值 ································ 159
　　　5.5.1　项目结构 ··· 159
　　　5.5.2　案例流程模块 ·· 159
　　　5.5.3　定义全局变量类 ·· 160
　　　5.5.4　数据库操作辅助类 ·· 161
　　　5.5.5　替换 area 表数据 ·· 167
　　　5.5.6　替换 school 表数据 ·· 170
　5.6　MySQL 数据库的构建 ·· 172
　　　5.6.1　MySQL 的应用范围 ·· 172
　　　5.6.2　MySQL 的优缺点 ·· 172
　　　5.6.3　MySQL 安装 ·· 173
　5.7　数据库可视化工具安装 ··· 179
　　　5.7.1　Navicat for MySQL 下载 ······································· 180
　　　5.7.2　Navicat for MySQL 安装 ······································· 181
　5.8　数据库表设计与数据的导入 ··· 184
　　　5.8.1　各表的结构设计 ·· 184
　　　5.8.2　建库和建表操作 ·· 191
　　　5.8.3　使用 SQL 语句建表 ··· 193
　　　5.8.4　SQL 语句讲解 ·· 197
　　　5.8.5　将 Excel 表导入数据库 ······································· 200
　　　5.8.6　将 SQL 文件导入数据库 ······································· 203
　5.9　本章小结 ·· 206
　5.10　课后练习 ··· 206

第 6 章　Java Web 服务器端设计　<<< 207

- 6.1 本章简介 ······ 207
- 6.2 服务器开发背景知识 ······ 207
  - 6.2.1 JSP 简介 ······ 207
  - 6.2.2 Tomcat 服务器 ······ 207
  - 6.2.3 服务器与客户端交互 ······ 209
- 6.3 了解 Java Web 技术 ······ 212
  - 6.3.1 DAO 设计模式 ······ 212
  - 6.3.2 认识 Java Web 程序的目录结构 ······ 213
- 6.4 Java Web 核心技术 ······ 213
  - 6.4.1 JavaBean 技术 ······ 213
  - 6.4.2 运行第一个 Java Web 程序 ······ 214
  - 6.4.3 Servlet 技术 ······ 217
  - 6.4.4 HttpServletRequest 类 ······ 217
  - 6.4.5 HttpServletResponse 类 ······ 218
  - 6.4.6 实战演练 ······ 218
- 6.5 设计 App 服务器数据库工具类 ······ 220
  - 6.5.1 JDBC 技术 ······ 220
  - 6.5.2 数据库连接类的实现 ······ 221
- 6.6 设计 App 服务器业务逻辑类 ······ 225
  - 6.6.1 建立实体类 ······ 225
  - 6.6.2 数据库操作类的实现 ······ 226
  - 6.6.3 练习 ······ 234
- 6.7 设计 App 服务器 Servlet 类 ······ 234
  - 6.7.1 省控线 Servlet 类的实现 ······ 235
  - 6.7.2 练习 ······ 239
- 6.8 App 服务器端设计巩固 ······ 240
  - 6.8.1 问题模块实体类 ······ 240
  - 6.8.2 问题模块数据库操作类 ······ 242
  - 6.8.3 问题模块 Servlet 类 ······ 248
- 6.9 本章小结 ······ 251
- 6.10 课后练习 ······ 251

第 7 章　App 客户端与服务器交互设计　<<< 252

- 7.1 本章简介 ······ 252
- 7.2 客户端和服务器端数据交互基础 ······ 253
  - 7.2.1 HttpClient ······ 253
  - 7.2.2 JSON 解析 ······ 255

        7.2.3　第三方 JAR 包导入 ································· 255
        7.2.4　客户端与服务器端交互工具类设计 ·············· 257
    7.3　"省控线查询"模块与服务器端交互的实现 ················· 262
        7.3.1　"省控线查询"模块与服务器端交互流程 ········ 262
        7.3.2　获取服务器端数据 ································· 262
        7.3.3　显示省控线数据列表 ······························ 264
    7.4　"历年录取线查询"模块与服务器端交互的实现 ············ 266
        7.4.1　"历年录取线查询"模块与服务器端交互流程 ··· 266
        7.4.2　获取服务器端学校录取线和专业录取线 ·········· 267
        7.4.3　显示学校录取线和专业录取线列表 ··············· 268
    7.5　"报考咨询"模块与服务器端交互的实现 ····················· 270
        7.5.1　"报考咨询"模块与服务器端交互流程 ············ 270
        7.5.2　获取历史问题列表 ································· 270
        7.5.3　实现"提问"模块 ··································· 273
        7.5.4　实现"问题回复"模块 ····························· 275
        7.5.5　实现"查询问题"模块 ····························· 278
    7.6　本章小结 ··· 278
    7.7　课后练习 ··· 279

第 8 章　Spring Boot 服务器端设计　　<<< 280
    8.1　本章简介 ··· 280
    8.2　Spring Boot 开发基础 ·· 280
        8.2.1　Spring Boot 技术简介 ····························· 280
        8.2.2　Spring Boot 项目开发环境 ······················· 281
        8.2.3　Spring Boot 项目开发基本过程 ·················· 284
    8.3　App 服务器实体层设计与实现 ································· 292
        8.3.1　App 服务器实体层设计 ···························· 292
        8.3.2　App 服务器实体层实现 ···························· 292
    8.4　App 服务器数据持久层设计与实现 ··························· 294
        8.4.1　App 服务器数据持久层设计 ······················ 294
        8.4.2　App 服务器数据持久层实现 ······················ 294
    8.5　App 服务器业务逻辑层设计与实现 ··························· 299
        8.5.1　App 服务器业务逻辑层设计 ······················ 299
        8.5.2　App 服务器业务逻辑层实现 ······················ 300
    8.6　App 服务器控制层设计与实现 ································· 302
        8.6.1　App 服务器控制层设计 ···························· 302
        8.6.2　App 服务器控制层实现 ···························· 303
    8.7　本章小结 ··· 304

8.8 课后练习 ··· 305

第 9 章　App 微信分享的实现　　<<< 306

9.1 本章简介 ··· 306
9.2 App 微信分享的操作流程 ··· 306
　　9.2.1 微信开放平台 ··· 306
　　9.2.2 将 App 内容分享给微信好友 ··· 307
　　9.2.3 将 App 内容分享到微信朋友圈 ··· 308
9.3 Android 应用打包签名 ·· 308
　　9.3.1 打包签名 apk 文件 ·· 309
　　9.3.2 配置 gradle 让 App 自动签名 ··· 311
9.4 Android 平台分享到微信的开发流程 ··· 312
　　9.4.1 申请 AppID ·· 312
　　9.4.2 搭建开发环境 ··· 314
　　9.4.3 实现微信分享功能 ··· 320
9.5 本章小结 ··· 321
9.6 课后练习 ··· 321

第 1 章

App 全栈开发概述

在本章中,我们开门见山地迈入移动开发的大门,通过"豹考通"App 介绍移动商务软件开发的知识模块与流程,并且从功能和界面的角度阐述"豹考通"从设计到开发的全过程。在此基础上,我们从高校教师教学的形式出发,以项目案例驱动形式辅助教学,摒弃以往的知识点讲解模式,提高学习效果。

1.1 什么是 App

App 是什么?有些读者可能有些了解,简单地说,App 是英文单词 Application 的简称,即智能手机的应用程序(无须连接 PC,直接运行在手机上的应用)。App 开发对于企业来说,需要根据自身的具体发展情况和战略规划来权衡。在一个好企业中,App 移动端开发一定是具有长期价值的一个重要组成部分。

App 拥有以下几个特点:
- App 安装在移动设备上,使用方便,也便于营销。
- App 融入了品牌元素,是品牌的一个具体体现,用户可以享受品牌的相关服务和最新资讯。
- App 与互联网相连,随时更新数据,提高用户体验。

1.2 App 开发流程

一个完整的软件开发分为网页端与移动端(App),而移动端又包括 Android 端和 iOS 端,使用的开发模式是移动应用开发常用的 MVC 开发模式。本书仅涉及 Android 端的开发。

1. 服务器框架搭建

开发流程如图 1-1 所示。

服务器端设计在一个移动项目设计中是很重要的部分,如果说客户端负责为用户展现数据,那么服务器端可以说是为用户准备数据的。只有服务器端将用户所需的数据准备完毕后,客户端才有可能去展示。

2. 数据库设计与部署

对于移动 App 开发,数据库的选择显得至关重要,因为数据存储结构和读取速度直接影响用户体验,所以要尽量使用轻量级数据库,这里我们使用的是 MySQL。

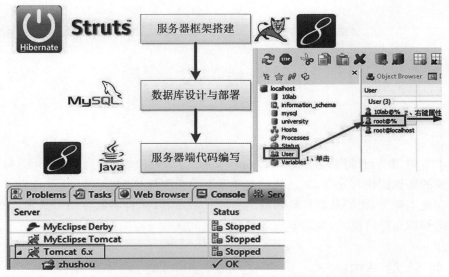

图 1-1 服务器端与数据库开发流程简图

数据库开发分为逻辑结构设计和物理结构设计。逻辑结构设计是对数据库实体属性的确定、E-R 图的确定以及对数据库存储结构进行设计,需要确定项目中所用到的字段并确定数据类型。物理结构设计就是在 MySQL 中创建相应的数据库表,对数据进行录入存储。最后需要将数据库与我们的后台服务器相连接。

3. 服务器端实现

服务器端是在 MyEclipse 2015 中实现的,首先要进行开发环境的配置。配置完成后,在 MyEclipse 2015 中创建"豹考通"工程。

4. 客户端框架搭建

客户端开发流程简图如图 1-2 所示。

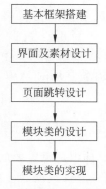

图 1-2 客户端开发流程简图

图 1-3～图 1-6 为开发时的主要工作界面。

第 1 章　App 全栈开发概述

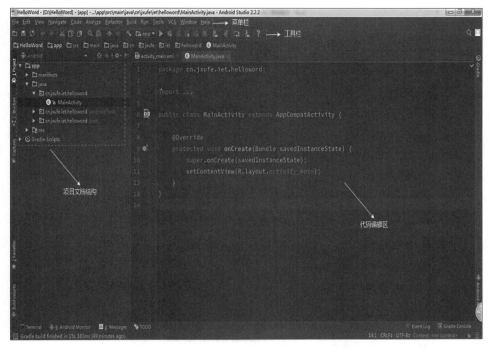

图 1-3　用 Android Studio 开发客户端主界面

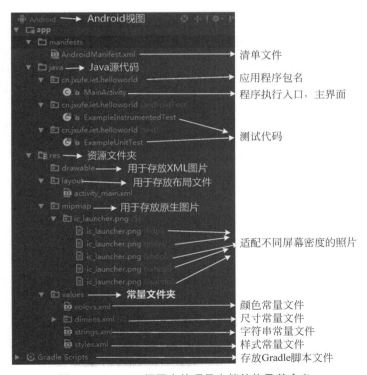

图 1-4　Android 视图中的项目文档结构及其含义

图 1-5　Project 视图中的项目文档结构及其含义

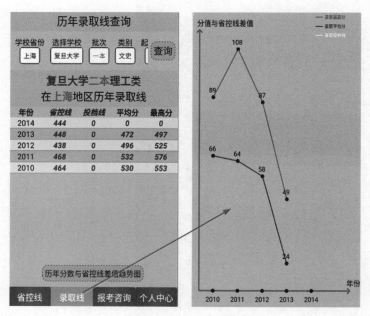

图 1-6　客户端页面跳转

5. 界面及素材设计和界面跳转设计

对于界面设计与模块类实现，可以分工同时进行。

项目素材由美工组制作完成，需要按照 Android 开发图标要求制作。在开发前，项目负责人需要指定一个详细的设计文档，指出需要的素材和规格。而界面与界面跳转设计，也需要通过讨论制订出想要的界面设计文档。其中需要注意的是，界面跳转不宜过多，除了必须跳转的界面外，可以在当前界面中实现的功能尽量在当前页面实现，这也是影响用户体验的重要指标之一。

6. 模块类的实现

根据项目的需求分析文档中列出的功能模块，在项目中分别创建模块文件夹，然后分别对功能模块进行设计与实现。

7. 网络数据读取接口

网络数据读取接口由服务器端提供，包括网址和需要的参数，客户端发送相应的参数请求从服务器端获取编译结果，最后由客户端呈现在用户眼前。网络数据接口形式如下所示。

```
http://localhost/zhushou/RequestControlLinesAction.action?c=0&y=2013&s=14&b=0
```

8. 编译打包及测试

移动应用编译打包、测试及发布上线的简易流程如图 1-7 所示。

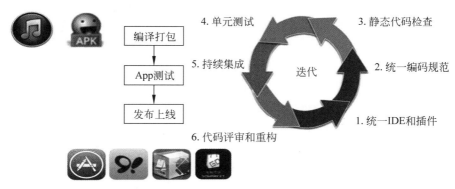

图 1-7　移动应用编译打包、测试及发布上线流程简图

对于 Android 端应用，需要在项目中打包生成 apk 文件；编译打包完成后，可以在其他成员的手机上安装测试，查看是否有问题。如果有，立即对程序进行调试；如果没有，则可以在商店中发布上线。

9. 发布上线

相对 iOS 发布过程来说，Android 应用的发布比较简单；而 iOS 应用的发布审核相对来说更为严格，一些细节问题的出现会让 Apple 公司拒绝。如果是发布在第三方商店（如"应用宝"），则审核相对顺利。

1.3 案例介绍

"豹考通"软件是一款面向全国高考学子和各高等院校教师的软件。高考志愿填报对考生录取高校与专业乃至未来的人生道路非常关键。由于信息不对称,考生填报志愿时比较盲目。"豹考通"借助智能手机,帮助考生在填报志愿时掌握全面信息,提供数据跟踪记录和深度分析,供考生参考,同时在高校与考生之间搭建一条新的交流途径。

此 App 已经在各 App 商店上架,读者可以先下载使用。

根据市场和往年高考情况综合考虑,"豹考通"实现了以下 7 个功能。

① 省控线查询:用户可以查询各省市高考省控线信息,使用方便,不需要到网上去搜集信息。

② 投档线查询:用户可以查询全国各高等院校的投档线信息。

③ 生成投档线曲线图:根据用户的信息选择,系统会为用户生成近几年的投档线和省控线的一个曲线图,更加直观地显示分数的变化波动情况,供用户参考。

④ 预测投档线:根据往年投档线情况,采用系统的算法,根据用户不同的选择,预测今年各学校投档线情况。

⑤ 推荐学校:推荐学校可以为用户推荐一些感兴趣并且有机会投档的院校,帮助用户做出报考选择。

⑥ 生成预测推荐报告:报告根据用户选择的参数生成,报告的内容包括今年投档线的预测和推荐的所有有机会投档大学的名单,为用户的最终选择提供参考。

⑦ 关于我们:介绍参与开发"豹考通"软件的团队成员。

"豹考通"Android 客户端的主要界面见第 3 章体验部分。

案例中访问服务器的部分均使用 www.xs360.cn 作为服务器地址,所有程序、软件包及源代码都可以从此网站中获取,当然读者也可以自己搭建本地服务器(访问局域网)。图 1-8 为客户端服务器网络连接示意图。

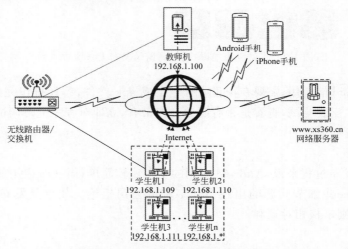

图 1-8 客户端服务器网络连接示意图

1.4 本章小结

本章主要带领大家进入 App 开发世界,介绍什么是 App;分析 App 整个项目从开发到发布的流程;另外介绍"豹考通"案例的主要功能,为后面结合案例学习 App 项目开发做好铺垫。

1.5 课后练习

1. 移动 App 开发一般包括哪几部分?试简要描述每个部分的流程。
2. 下载"豹考通"App,思考其界面跳转设计的思路并提出自己的见解。
3. 搭建本地服务器并在网页中试用 Web 版应用。

项目需求分析

2.1 本章简介

需求分析是产品研发前期的铺垫工作,也是重要的基础工作之一。需求分析工作中的缺陷将给项目带来极大风险,在质量、功能、场景等情境下影响用户的满意度和期望值。因此,我们在实现"豹考通"项目之前,需要对项目进行需求分析。由于本书介绍"豹考通"的实现,因此本章从功能需求分析、系统用例分析、数据库分析三个方面对"豹考通"项目进行简单分析,为后面设计和开发阶段提供基础。

2.2 功能需求分析

随着移动互联网技术的不断发展,一些方便用户日常生活的查询工具也应运而生。我们的"豹考通"应用就是面向广大高考学生提供服务,目的在于让他们能够更加方便地查询到近几年与高考相关的数据。

2.2.1 需求获取

在开发软件之前需要通过和用户的交流以及和团队之间的讨论来确定软件需求,使系统更加详尽和准确。下面是需求获取过程示意图,如图 2-1 所示。

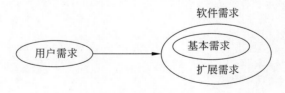

图 2-1 需求获取过程

一个项目的需求可分为两个部分:一是基本需求,二是扩展需求。扩展需求是暂时不能实现的需求部分,但是作为系统的一部分会在以后的工作中加以完善。

2.2.2 功能分析

下面我们对"豹考通"软件的功能需求进行相关的分析。

1. 注册登录

应用为考生提供注册登录的功能,考生在注册时只需要填写姓名、邮箱和密码即可,邮箱作为必填选项。在"生成报考单"功能中,应用会将最后的报告单以邮件的形式发送到考生的注册邮箱中。这就使得考生在使用"生成报告单"功能时,需要先登录,系统在读取用户信息之后才能将报告单发送到考生邮箱。

2. 查询省控线

省控线指的是省(区、市)划定的相应本科、专科批次的最低录取控制分数线。它一般用于以下情形。

① 某些学校没招满或扩招,降低分数线来录取。

② 某些考生单科成绩极其突出,降低分数线来录取。

"豹考通"为考生提供了查询省控线的功能,这也是作为一个高考考生比较关心的问题,方便了考生查找信息。

3. 查询投档线

投档线即投档分数线,院校投档分数线是指以院校为单位,按招生院校同一科类(如文科或理科)招生计划数的一定比例(1:1.2 以内),在对第一志愿投档过程中自然形成的院校调档最低成绩标准。每一所院校都有自己的投档线,投档线也称调档线或提档线。

学校历年的投档线也是考生关注的一个重要信息,"豹考通"提供了全国各高校近几年的投档线信息并给出了最低分和平均分信息,还根据省控线和这些分数的差值画出了相应的分数差值曲线,便于学生参考。

4. 推荐学校

高考之后选择学校可能是让考生最头疼的事情,自己感兴趣的学校可能分数又不够,那么如何最合理地根据自己的高考成绩选择学校呢?

"豹考通"为考生提供了一个推荐学校的功能,该功能分为两个子功能:一是根据考生的高考成绩为考生推荐学校,而且可以选择三种不同的推荐意向方式(冒险、保守和稳妥),这三种意向代表着推荐的风险大小,选择稳妥可能会使推荐的学校更有机会录取;二是根据考生的省排名信息和意向专业来推荐,但前提是考生必须完善自己的个人信息,如排名、生源地、专业意向和意向省市等。同样,也可以选择三种不同的推荐意向方式,从冒险到稳妥,其风险越来越小。

5. 预测当年投档线

"豹考通"已经为考生提供了查询投档线的功能,在此基础上,根据得到的历年数据对一些算法加以改进,可预测当年投档线。考生输入学校信息和今年的省控线信息后,系统可以根据相关信息预测出该学校当年的投档线,为考生提供一个参考。

6. 生成报告单

报告单是基于考生高考成绩、省排名、生源地、兴趣专业等信息为考生提供的一份详细的报考指南,它列出了有可能被录取而且是考生感兴趣的学校,并且列出了该学校的相关专业信息。已经登录的考生可以通过注册时填写的邮箱地址收到报告单。

7. 联系各院校招生单位

有的考生在选择好一所自己感兴趣的学校后,可能还需要上官方网站查询各个专业的信息、负责人信息等,"豹考通"提供了各院校招生单位的联系方式,这样考生不必上网查询,直接打开应用,里面就有相关信息,还能直接拨打招生单位电话,和联系人直接交流。

2.2.3 系统 UML 建模

包是一种组合机制,帮助用户了解系统中相关元素之间的关系,形成一个高内聚、低耦合的整体。

项目整体结构包图如图 2-2 所示。

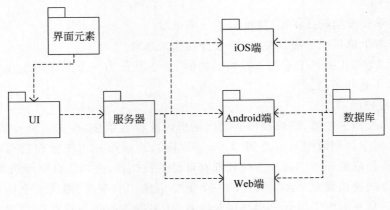

图 2-2 项目整体结构包图

2.3 系统用例分析

由于本书的重点在于 App 开发过程的实现部分,因此在用例分析部分只是简单地给出系统的用例图等,供读者参考。

2.3.1 系统用例图

系统用例图如图 2-3 所示。

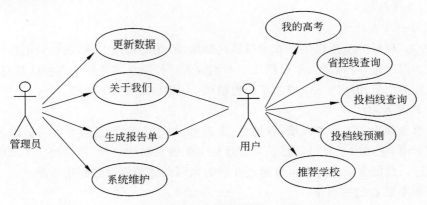

图 2-3 系统用例图

2.3.2 系统用例描述

用户前台用例分析如表 2-1～表 2-6 所示。

表 2-1 查询省控线用例分析

用例名称：查询省控线
用例标识号：102
参与者：用户
简要说明：用户输入相关参数查询近 8 年全国各省省控线
前置条件： 1. 用户联入网络 2. 输入相应参数（省份、年份、类别、批次）
基本事件流： 1. 用户打开手机客户端或 Web 端，选择查询省控线功能 2. 输入相应参数（省份、年份、类别、批次） 3. 单击"查询"按钮，系统显示相关省控线数据 4. 用例终止
其他事件流 A1：无
异常事件流： 1. 如果输入的参数没有数据，系统会弹出提示对话框，提示用户没有相应的数据 2. 如果用户没有联网，系统会提示用户先联网再查询信息
后置条件：无
注释：无

表 2-2 查询投档线用例分析

用例名称：查询投档线
用例标识号：103
参与者：用户
简要说明：用户输入相关参数查询近 8 年全国各院校投档线
前置条件： 1. 用户联入网络 2. 输入相应参数（省份、学校、年份、类别、批次）
基本事件流： 1. 用户打开手机端或 Web 端，选择查询投档线功能 2. 输入相应参数（省份、学校、年份、类别、批次） 3. 单击"查询"按钮，系统显示相关投档线数据 4. 用例终止
其他事件流 A1：无

续表

用例名称：查询投档线

异常事件流：
1. 如果输入的参数没有数据，系统会弹出提示对话框，提示用户没有相应的数据
2. 如果用户没有联网，系统会提示用户先联网再查询信息

后置条件：无

注释：无

表 2-3 预测投档线用例分析

用例名称：预测投档线

用例标识号：104

参与者：用户

简要说明：用户输入相关参数，根据算法预测本年学校的投档线

前置条件：
1. 用户联入网络
2. 输入相应参数（省份、学校、类别、批次）

基本事件流：
1. 用户打开手机端或 Web 端，选择预测投档线功能
2. 输入相应参数（省份、学校、类别、批次）
3. 单击"预测"按钮，系统显示本年预测投档线结果，结果有三种不同倾向：冒险、保守和稳妥
4. 用例终止

其他事件流 A1：无

异常事件流：
3. 如果输入的参数没有数据，系统会弹出提示对话框，提示用户没有相应的数据
4. 如果用户没有联网，系统会提示用户先联网再查询信息

后置条件：无

注释：无

表 2-4 推荐学校用例分析

用例名称：推荐学校

用例标识号：105

参与者：用户

简要说明：用户输入相关参数为用户推荐有可能被录取的学校

前置条件：
1. 用户联入网络
2. 输入相应参数（生源地、类别、分数、批次、预测倾向、目标省份）

续表

用例名称：推荐学校

基本事件流：
1. 用户打开手机端或 Web 端,选择推荐学校功能
2. 输入相应参数(生源地、类别、分数、批次、预测倾向、目标省市)
3. 单击"推荐"按钮,系统显示相关学校的信息
4. 用例结束

其他事件流 A1：无

异常事件流：
5. 如果输入的参数没有数据,系统会弹出提示对话框,提示用户没有相应的数据
6. 如果用户没有联网,系统会提示用户先联网再查询信息

后置条件：无

注释：无

表 2-5　生成报告单用例分析

用例名称：生成报告单

用例标识号：106

参与者：用户、管理员

简要说明：用户输入相关参数为用户生成一份报考志愿报告单,管理员通过用户所留邮箱地址将报告单发送到用户邮箱

前置条件：
1. 用户联入网络
2. 用户已经登录
3. 输入相应参数(生源地、类别、分数、批次、目标省份、兴趣专业)

基本事件流：
1. 用户打开手机端或 Web 端,选择生成报告单功能
2. 输入相应参数(生源地、类别、分数、批次、目标省份、兴趣专业)
3. 单击"生成报告单"按钮,管理员将报告单以邮件的形式发送给用户
4. 管理员登录系统,通过用户所留邮箱将报告单发送给用户
5. 用例终止

其他事件流 A1：无

异常事件流：
1. 如果输入的参数没有数据,系统会弹出提示对话框,提示用户没有相应的数据
2. 如果用户没有联网,系统会提示用户先联网再生成报告
3. 如果用户没有登录(注册),系统会提示用户先登录(注册),再生成报告

后置条件：无

注释：无

表 2-6　关于我们用例分析

用例名称：关于我们
用例标识号：107
参与者：用户
简要说明：用户可以查看与软件开发者相关的信息
前置条件：无
基本事件流： 1. 用户打开手机端或 Web 端，选择"关于我们"功能 2. 用例终止
其他事件流 A1：无
异常事件流：无
后置条件：无
注释：无

管理员后台用例分析如表 2-7 和表 2-8 所示。

表 2-7　添加相关数据用例分析

用例名称：添加相关数据
用例标识号：108
参与者：管理员
简要说明：管理员根据网上已经更新的数据，将数据添加到系统中，供用户使用
前置条件：管理员登录系统
基本事件流： 1. 管理员将最新的数据添加到系统后台数据库中 2. 用例终止
其他事件流 A1：无
异常事件流：无
后置条件： 用户在前台可以查询最新的相关信息
注释：无

表 2-8　发送报告单用例分析

用例名称：发送报告单
用例标识号：109
参与者：管理员
简要说明：用户生成报告单之后，管理员根据用户邮箱将报告单发送给用户

续表

用例名称：发送报告单
前置条件： 1. 管理员登录系统 2. 用户填写了邮箱信息
基本事件流： 1. 管理员将系统生成的报告单发送到用户邮箱 2. 用例终止
其他事件流 A1：无
异常事件流： 如果用户没有填写邮箱信息，系统会提醒用户完善自己的信息
后置条件： 用户可以登录邮箱查看自己的报告单
注释：无

2.4 项目数据库分析

2.4.1 数据库基本概念

数据库，顾名思义就是应用中所需的数据都保存在其中，数据库设计的简单与复杂直接影响到应用设计的框架和结构。因此，在进行数据库设计时，需要用户花费比较多的时间去思考应用中所需的数据并将它们整理成数据表。

在人们平常所提到的数据处理概念中，最主要的就是"数据"和"信息"，那么它们各自代表什么含义呢？信息是对现实世界中存在的客观实体、现象和关系进行描述的具有特定意义的数据，是经过加工处理的数据。

- 数据：从字面上理解数据就是用来描述客观事物的符号、标记、图形等信息的组合；从实际应用上来看，数据能够直观地体现出与应用相关的信息。
- 信息：以数据为载体的对客观实际存在的事物、事件和概念的抽象反映，是经过加工处理的数据。
- 数据处理：数据处理就是将数据进行相应的加工，处理之后成为信息。数据处理是将数据转换成信息的过程，是对信息进行收集、整理、存储、加工以及传播等一系列活动的总和。

数据和信息二者之间的关系是密不可分的，我们可以简单地运用下面的公式表述数据、信息和数据处理三者之间的关系。

$$信息 = 数据 + 数据处理$$

数据是原材料，而信息则是产品，数据处理的真正意义是为了产生信息而去处理相关的数据。

2.4.2 数据库设计的基本过程

数据库设计的过程主要是三个阶段：需求分析阶段、数据库设计与实施阶段和数据库维护阶段。

1. 需求分析阶段

简单来说，需求分析阶段就是通过详细周全的调查和分析总结出系统中所需完成的任务。需求分析阶段是整个数据库设计的开始，如果需求分析做得不好，那么必定会对后面的数据库设计阶段造成很多麻烦，增加工作量。

2. 数据库设计与实施阶段

数据库设计阶段就是将需求分析阶段总结出的一些任务需求细化到每个字段，通过字段表示出应用系统中需要的数据信息并设计出相应的数据表，列出每个表的主键、外键等并设计每个数据库表之间的联系，最后存入数据库中，便于后面应用中调用。设计阶段也是整个数据库设计中最为重要的一个阶段。

3. 数据库维护阶段

数据库维护阶段是指在数据库设计与实施阶段完成之后，对数据库的数据进行更新，同时也会根据需求的变动来更改数据库的相关字段。不过我们不提倡在确定需求之后再去对应用系统进行较大的修改。

也可以将数据库设计过程划分得更详细，一种常见的方法是将之分成需求分析、概念设计、逻辑设计、物理设计、数据库实施、数据库运行和维护六个阶段。

2.4.3 系统涉及的实体及其属性

"生源省市"实体主要保存我国各个省或直辖市的信息，如图2-4所示。"学校所在地区"实体主要保存学校所在省市县的信息，如图2-5所示。

图2-4 "生源省市"实体图　　　　图2-5 "学校所在地区"实体图

"批次"实体主要保存录取的批次信息，包括一本、二本、三本、专科和提前本科五种，如图2-6所示。"科类"①主要保存专业类别，分为文史、理工、综合、体育和艺术五类，如图2-7所示。

图2-6 "批次"实体图　　　　图2-7 "科类"实体图

---

① 本书与考试科类有关的数或字段名称中，"科类"与"类别"指同一含义。

"学校"实体的主要属性包括学校编号、学校代码、学校名称、是否211、是否985、是否教育部直属以及学校所在地区,如图2-8所示。

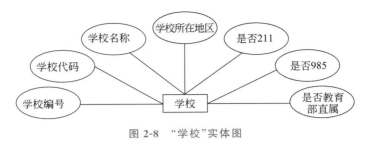

图 2-8 "学校"实体图

学校与学校所在地区为一对多的关系,如图2-9所示。一个地区可以有多个学校,但一个学校只位于一个地区,当学校在其他地区有分校时,将其看作两个学校。

图 2-9 学校与学校所在地区 E-R 图

每个学校有一个学校招生部门。"学校招生部门"实体包括学校联系信息编号、学校名称、用户名(用于登录)、密码、联系人姓名、邮箱、办公室电话、手机号码、QQ 号、网址、被联系次数、创建时间、最后登录时间等属性,如图 2-10 所示。

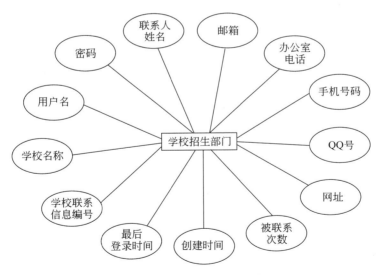

图 2-10 "学校招生部门"实体图

"招生部门联系人"实体主要包括联系人编号、联系人姓名、职位、电话和邮箱等属性,如图 2-11 所示。

"专业"实体主要包括专业编号、专业名称、专业代码等属性,如图 2-12 所示。

学校的每个专业有一个专业负责人,"专业负责人"实体主要包括负责人编号、负责人

图 2-11 "招生部门联系人"实体图

图 2-12 "专业"实体图

姓名、负责人电话和负责人邮箱等属性,如图 2-13 所示。

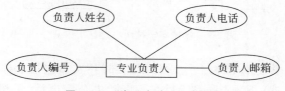

图 2-13 "专业负责人"实体图

学校、专业及专业负责人有关系。一个学校开办多个专业,一个专业可以由多个学校开办,因此学校和专业是一对多的关系。一个专业拥有一个专业负责人,一个专业负责人负责一个专业,因此专业和专业负责人是一对一的关系。学校与专业 E-R 图如图 2-14 所示。

图 2-14 学校与专业 E-R 图

生源省市与批次、科类等实体有关系,同时还应包括年份、分数和省控线编号属性。其中生源省市与批次、科类的关系均为一对多,只有批次、科类以及年份三个因素才能唯一确定一个生源省市的省控线分数,如图 2-15 所示。

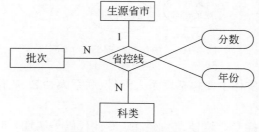

图 2-15 省控线 E-R 图

学校与生源省市、批次、科类、学校所在地区等实体有关系,同时还应包括年份、平均分、投档线、最高分属性。其中学校与批次、生源省市、科类间的关系均为一对多;学校与学校所在地区是一对一的关系。只有生源省市、批次、科类以及年份四个因素才能唯一确定此学校某年具体批次具体科类的录取线。录取线包括平均分和最高分两项,如图 2-16 所示。

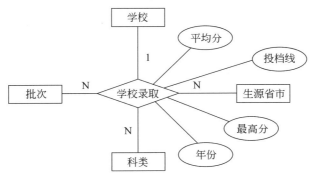

图 2-16　学校录取 E-R 图

学校与专业、生源省市、批次、科类等实体有关系,同时还应包括年份、省控线、最高分、平均分属性。其中专业与学校、生源省市与学校、批次与学校、科类与学校间的关系均为一对多,只有专业、生源省市、批次、科类以及年份五个因素才能唯一确定此学校某个专业某年具体批次具体科类的录取线,如图 2-17 所示。录取线包括平均分和最高分两项。注意,同一个专业在不同学校批次可能不同,有些专业既可招文科生也可招理科生。

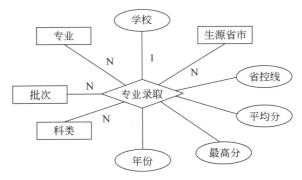

图 2-17　学校与专业录取 E-R 图

"高考咨询"实体保存与高考相关的咨询信息,包括咨询编号、生源省市编号和咨询内容,如图 2-18 所示。

图 2-18　"高考咨询"实体图

"豹考通"App 数据总体 E-R 图如图 2-19 所示。

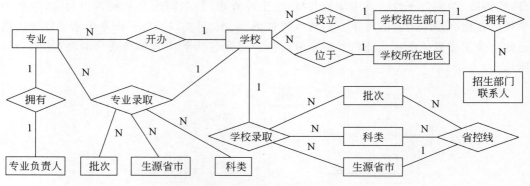

图 2-19 "豹考通"App 数据总体 E-R 图

2.5 本章小结

本章完成了设计和实现"豹考通"项目之前的需求分析工作。由于本书的主要任务放在 App 的设计和实现上,需求分析主要从 3 个方面进行展开:功能需求分析,根据需求得出项目的主要功能并进行 UML 建模;系统用例分析,给出系统的用例图,对用例进行详细的描述;项目数据库分析,介绍数据库的概念和数据库设计的基本过程,针对"豹考通"项目进行数据库设计,分析数据库涉及的实体及属性。

2.6 课后练习

打开"豹考通"App,对照本章 2.2 节内容,了解 App 开发的功能需求分析过程。

第 3 章

App 应用体验

3.1 本章简介

本章主要体验"豹考通"项目,介绍"豹考通"项目的功能、如何搭建开发环境、开发工具下载及安装过程和虚拟模拟器配置,讲解服务器端和客户端的运行过程以及 Android 应用目录结构文件的作用,达到体验"豹考通"功能的效果。为便于 Android 初学者学习,针对"豹考通"项目设计的两个版本是本地版"豹考通"和网络版"豹考通"。

"豹考通"是一款辅助高考学生填报志愿的应用软件,通过该软件,考生可以查询各省历年的省控线、各个学校及其专业在本省历年的录取线并可生成趋势图,同时软件提供查询学校和专业排名、模拟志愿填报与录取、报考咨询、性格测试推荐专业等服务。本书中的项目是真实的"豹考通"软件的简化教学版,重在讲解 App 开发流程和涉及的关键技术,对数据和功能进行了简化。它主要剖析了软件中的几个关键功能模块,详细讲解其开发流程和操作步骤,使读者对 App 开发流程有一个较全面的认识。如果读者对完整的"豹考通"项目感兴趣,可到应用宝、安智市场等平台下载完整版"豹考通"App。

本书主要讲解的功能模块有往年省控线查询、学校录取线查询、学校录取线趋势图、专业录取线查询、报考咨询、个人基本信息、本地数据库、主页面选项卡、远程数据库、服务器端、客户端与服务器端交互等。整体的功能结构如图 3-1 所示。

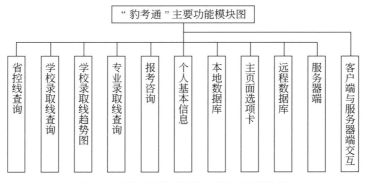

图 3-1 "豹考通"项目的主要功能模块图

3.2 开发环境配置

3.2.1 Java 语言

"豹考通"项目由 Android 客户端、服务器端和数据库三大部分组成。Android 客户端和服务器端基于 Java 语言进行开发。开发之前,我们需要配置 Java 开发环境。

1996 年,Sun 公司发布了 JDK 1.0 和 HotJava。通过嵌入网页中的 Applet 运行 Java 程序,这一特性相继被 Microsoft 公司和 Netscape 公司的 Web 浏览器所支持。IBM、Apple、DEC、Adobe、Silicon Graphics、HP、Oracle 和 Microsoft 等公司相继购买了 Java 技术许可证,从此 Java 成为日益流行的计算机语言。

JDK 是一个编写 Java 的 Applet 小程序和应用程序的程序开发环境。它是整个 Java 的核心,包括 Java 运行时环境(Java Runtime Environment,JRE)、一些 Java 工具和 Java 的核心类库(Java API)。无论什么 Java 应用服务器,实质都是内置了某个版本的 JDK。主流的 JDK 是 Sun 公司发布的 JDK,除 Sun 之外,还有很多公司和组织都开发了自己的 JDK。例如,IBM 公司开发的 JDK、BEA 公司的 JRocket、GNU 组织开发的 JDK。

另外,可以把 Java API 类库中的 Java SE API 子集和 Java 虚拟机这两部分统称为 JRE(Java Runtime Environment),JRE 是支持 Java 程序运行的标准环境。JRE 是运行环境,JDK 是开发环境。因此,写 Java 程序时需要 JDK,而运行 Java 程序时就需要 JRE。而 JDK 里面已经包含了 JRE,因此只要安装了 JDK,就可以编辑 Java 程序,也可以正常运行 Java 程序。但由于 JDK 包含了许多与运行无关的内容,占用的空间较大,因此运行普通的 Java 程序无须安装 JDK,而只需要安装 JRE 即可。

3.2.2 Java 环境配置

1. 下载 JDK

首先需要下载 Java 开发工具包 JDK,下载地址为 http://www.oracle.com/technetwork/java/javase/downloads/index.html,单击 DOWNLOAD 按钮,如图 3-2 所示。

在下载页面中你需要选择接受许可并根据自己的系统选择对应的版本,如果计算机操作系统为 64 位,可以安装 64 位 JDK,也可以安装 32 位 JDK。如果操作系统为 32 位,则只能安装 32 位 JDK(如图 3-3 所示)。但要注意,本书后面使用的工具 Eclipse 和 MyEclipse 的位数需要和 JDK 保持一致,本书以 Windows 64 位系统为例。

2. 安装文件

单击下载好的 JDK 安装包,会出现图 3-4 所示的安装界面,单击"下一步"按钮,如图 3-4 所示。

在如图 3-5 所示的 JDK 安装地址对话框中单击"更改"按钮,可随意更改 JDK 的安装目录。本文以修改为 D:\Java\jdk1.8.0_101\为例(见图 3-6)。

单击"下一步"按钮(见图 3-7),完成 JDK 安装(见图 3-8)。

在 JDK 安装过程中,系统会自动安装 JRE,单击"更改"按钮可进行安装目录更改(见

第 3 章　App 应用体验

图 3-2　JDK 下载按钮

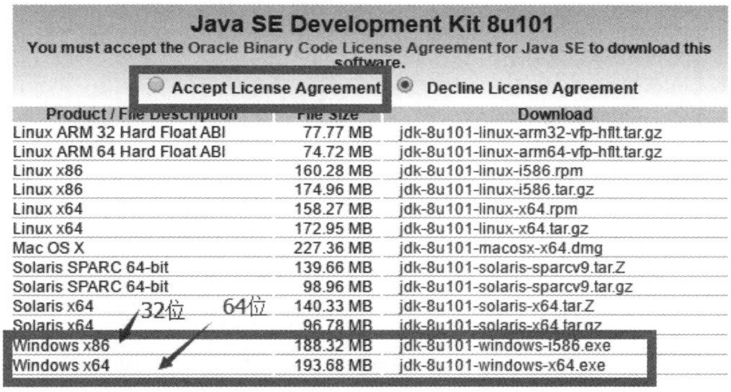

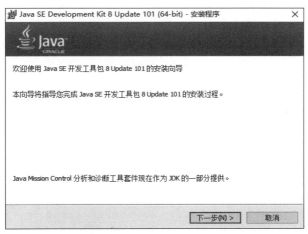

图 3-3　JDK 下载列表

图 3-4　JDK 安装引导页面

图 3-5　JDK 安装地址对话框

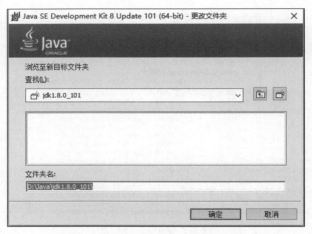

图 3-6　JDK 修改安装目录对话框

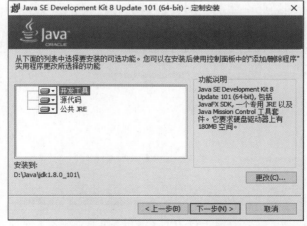

图 3-7　修改完成的 JDK 安装目录对话框

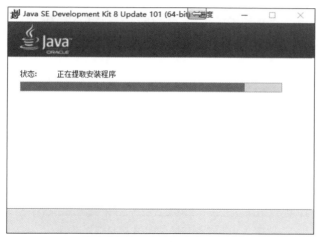

图 3-8　JDK 安装进度对话框

图 3-9)。建议 JRE 和 JDK 放在同一目录下，本文以 D:\Java\jre1.8.0_101 为例，在 D 盘 Java 文件夹下新建 jre1.8.0_101 文件夹，单击"更改"按钮后，在弹出的对话框中选择 jre1.8.0_101 文件夹完成目录更改(见图 3-10)。

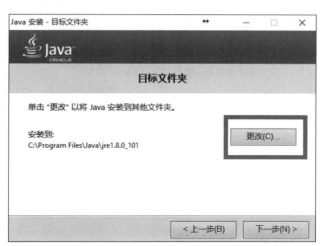

图 3-9　JRE 安装地址对话框

在图 3-10 上单击"下一步"按钮完成 JRE 安装，如图 3-11 和图 3-12 所示。

3. 配置 Java 环境

在 Java JDK 1.5 之前，Java JDK 安装完成后并不能立即使用，还需要配置相关环境变量。Java JDK 1.5 之后系统会有默认的配置，但建议手动进行配置。右击"我的电脑"，单击"属性"，选择"高级系统设置"选项，如图 3-13 所示。

单击"环境变量"按钮，如图 3-14 所示。

在系统变量中新建一个 JAVA_HOME 变量，选择"新建"按钮，如图 3-15 所示。

"变量名"输入 JAVA_HOME，"变量值"输入 D:\Java\jdk1.8.0_101(要根据自己

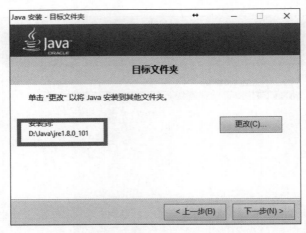

图 3-10　修改好的 JRE 安装目录对话框

图 3-11　JRE 安装进度对话框

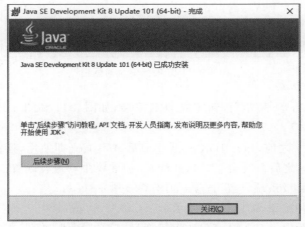

图 3-12　JDK 安装成功对话框

图 3-13 "系统"对话框

图 3-14 高级系统设置对话框

图 3-15 "环境变量"对话框

JDK 的实际路径配置),然后单击"确定"按钮,如图 3-16 所示。

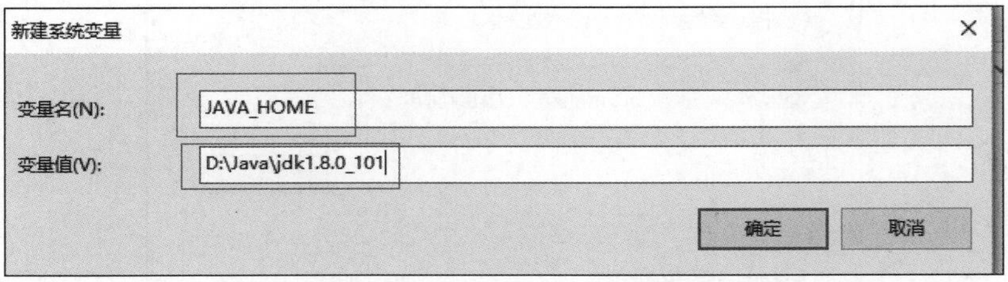

图 3-16 设置 JAVA_HOME 环境变量

建议 JAVA_HOME 变量名为大写,表示常量。但 Windows 系统不区分大小写,即大写、小写、大小写混合表示同一个变量名,虽不会出错,但不符合规范。

注意:变量值后不需要加任何符号。

选择 Path 选项,在变量值后面追加输入"％JAVA_HOME％\bin;％JAVA_HOME％\jre\bin;",％JAVA_HOME％代表的路径就是 D:\Java\jdk1.8.0_101\。

注意:输入时需要输入法是英文状态,然后单击"确定"按钮,如图 3-17 和图 3-18 所示。

单击"新建"按钮,"变量名"输入 CLASSPATH,"变量值"输入".;％JAVA_HOME％\lib\dt.jar;％JAVA_HOME％\lib\tools.jar;",其中点(.)表示当前目录,分号表示多个路径之间的分隔符,然后单击"确定"按钮(见图 3-19 和图 3-20)。

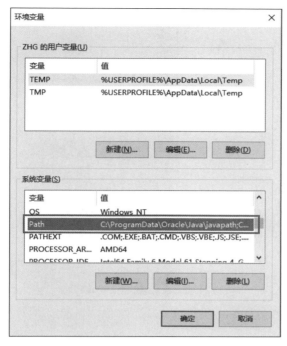

图 3-17　Path 路径

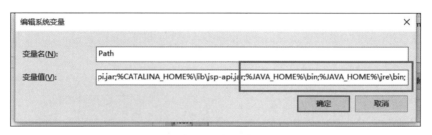

图 3-18　在 Path 变量中添加路径

后面连续单击"确定"按钮关闭对话框,这样 Java 环境就配置完成了。

4. 环境变量配置测试

（1）选择"开始"→"运行",输入 cmd。

（2）输入 java -version、java、javac 等命令,如果出现如图 3-21 所示的信息,则说明环境变量配置成功。

3.2.3　Android Studio 下载

Android Studio 是一个 Android 开发环境,基于 IntelliJ IDEA。类似 Eclipse ADT,它提供了集成的 Android 开发工具用于开发和调试。

Google 已宣布,为简化 Android 的开发力度以重点建设 Android Studio 工具,现在已经停止支持 Eclipse 等其他集成开发环境。而随着 Android studio 正式版的推出和完

图 3-19 "环境变量"对话框

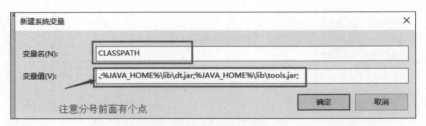

图 3-20 设置 CLASSPATH 环境变量

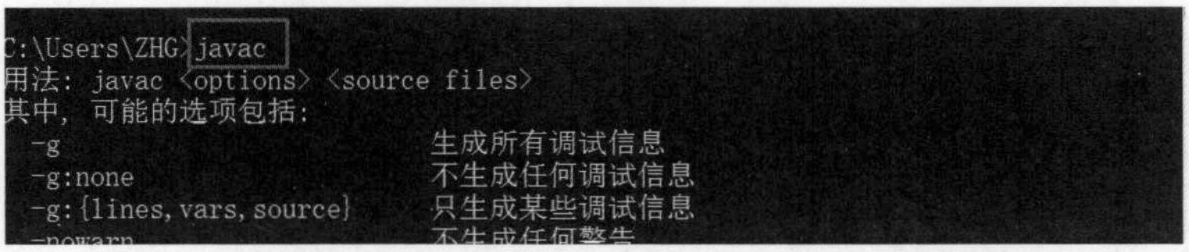

图 3-21 在命令行窗口中键入 javac 命令

善，Android 开发者大多已转向 Android Studio 开发平台。

下载 Android Studio 很简单，利用搜索引擎搜索 Android Studio 就可以轻松找到其最新版本的下载；也可以在浏览器中打开下面这个站点下载：

    developer.android.com/sdk/installing/studio.html

3.2.4　Android Studio 安装

（1）双击 Android Studio 的安装文件，进入安装界面，如图 3-22 所示。

图 3-22　Android studio 安装界面

（2）选择要安装的插件，如图 3-23 所示。

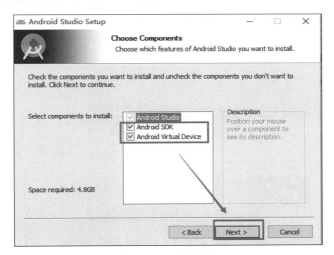

图 3-23　安装插件

第一个是 Android Studio 主程序，必选。第二个是 Android SDK，会安装 Android 的 SDK，也勾上。第三个是 Android Virtual Device，用于安装 Android 虚拟设备。

完成后单击 Next 按钮。

（3）同意条款后选择 I Agree 按钮，如图 3-24 所示。

（4）选择 Android Studio 和 SDK 的安装目录，如图 3-25 所示。

（5）设置虚拟机硬件加速器可使用的最大内存，如图 3-26 所示。

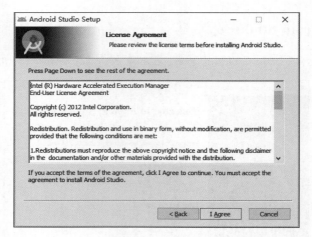

图 3-24　同意条款

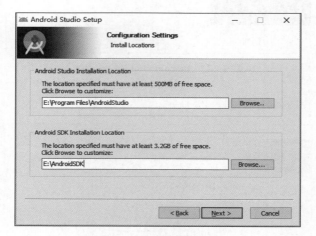

图 3-25　选择安装目录

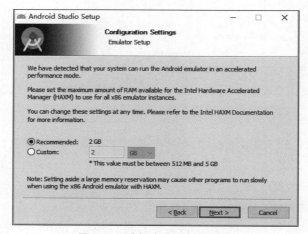

图 3-26　选择虚拟加速器内存

如果计算机配置还不错,默认设置 2GB 即可,如果配置较差,则选择 1GB,否则会影响运行其他软件。

(6) 之后就进入自动安装模式,如图 3-27 所示。

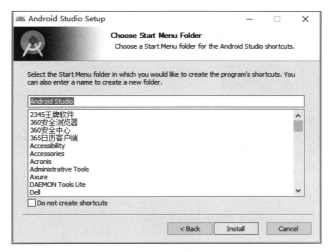

图 3-27　自动安装

如果没出什么意外,一小段时间后你就会看到如图 3-28 所示的界面,也就说明安装成功了,如图 3-29 所示。

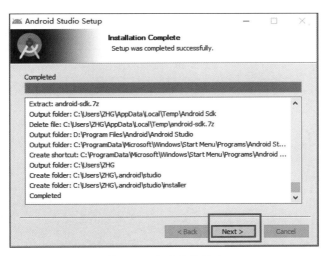

图 3-28　自动安装结束

(7) 打开 Android Studio 后,进入相关配置界面,如图 3-30 所示。

这用于导入 Android Studio 的配置文件,如果是第一次安装,选择最后一项,即不导入配置文件,然后单击 OK 按钮即可。

(8) 完成后就会进入如图 3-31 所示的页面,这是程序在检查 SDK 的更新情况,然后开始下载组件,如图 3-32 所示。

图 3-29　安装成功页面

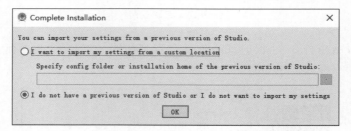

图 3-30　配置界面

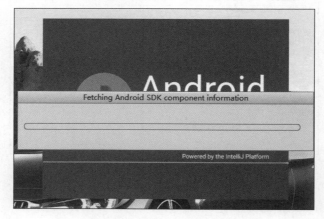

图 3-31　检查 SDK 更新情况

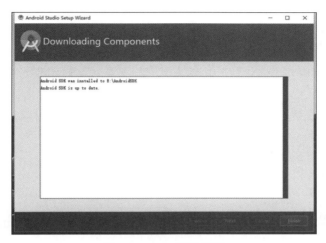

图 3-32　SDK 下载界面

3.2.5　创建第一个项目：HelloWorld

（1）创建 HelloWorld 项目。和其他平台类似，在这之前假设你已经配置好 JDK 和 Android SDK 环境，并且第一次安装 Android Studio。当更新完 Android SDK 后，将会看到如图 3-33 所示的界面。

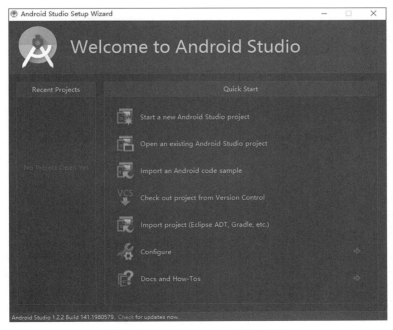

图 3-33　Android Studio 初始界面

- 选项 1：创建一个 Android Studio 项目。

- 选项2：打开一个 Android Studio 项目。
- 选项3：导入官方样例，从网络上下载代码。此功能在以前的测试版本中是没有的，建议查看官方给出的范例。
- 选项4：从版本控制系统中导入代码，支持 CVS、SVN、Git、Mercurial 以及 GitHub。
- 选项5：导入非 Android Studio 项目，例如原生的 IDEA 开发的项目。
- 选项6：设置。
- 选项7：帮助文档。

(2) 填写应用名和包名，如图 3-34 所示。

图 3-34　填写应用名和包名

(3) 选择安卓版本，如图 3-35 所示。

(4) 选择适应的界面，如图 3-36 所示。

在这个页面中选择一个 Activity 模板，和 Eclipse 一样，我们直接选择一个 Empty Activity。

(5) 确认 Activity Name 和 Layout Name 后完成创建，如图 3-37 所示。

至此，一个简单的 Android Studio 项目就完成了，如图 3-38 所示。

3.2.6　创建一个虚拟机设备

(1) 单击菜单栏 AVD Manager 按钮后，弹出对话框，在对话框中单击 Create Virtual

第 3 章　App 应用体验

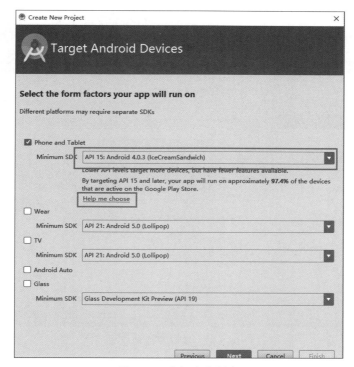

图 3-35　选择安卓版本

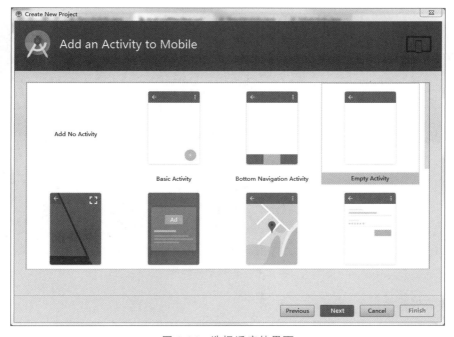

图 3-36　选择适应的界面

图 3-37　确认完成创建

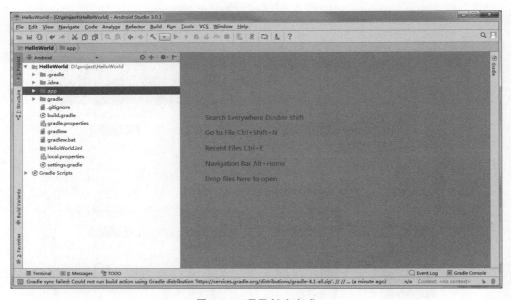

图 3-38　项目创建完成

Device 按钮创建模拟器,如图 3-39 所示。

(2) 在 Category 栏中选择 Phone,然后选择合适的屏幕尺寸和分辨率,单击 Next 按钮,如图 3-40 所示。

(3) 选择合适的镜像文件。如果没有,可单击 Download 按钮下载对应镜像文件(见

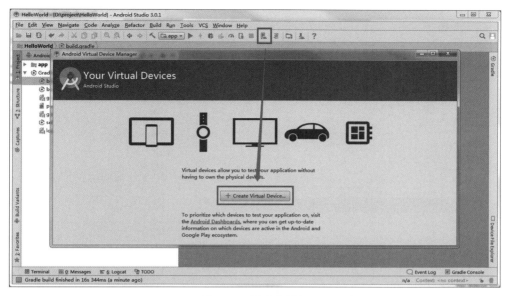

图 3-39　创建虚拟机

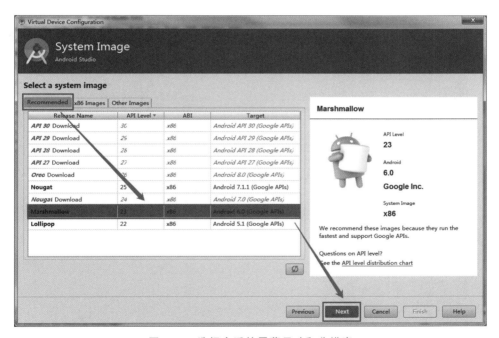

图 3-40　选择合适的屏幕尺寸和分辨率

图 3-41）。

（4）输入虚拟模拟器名，然后单击 Finish 按钮，这样虚拟模拟器就创建完成，如图 3-42 所示。

（5）单击启动按钮，启动模拟器，如图 3-43 所示。

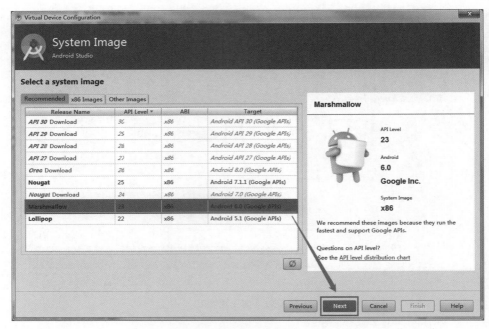

图 3-41　选择合适镜像文件

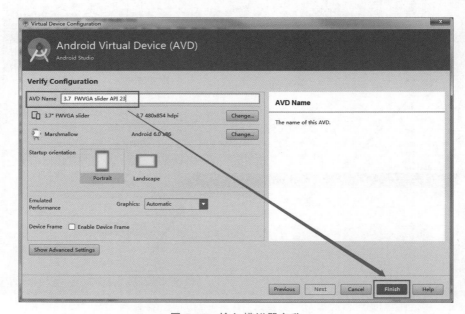

图 3-42　输入模拟器名称

（6）单击导航栏运行按钮，在模拟器中运行项目，效果如图 3-44 所示。

提示：

（1）在 Android Studio 中启动一个模拟器去运行程序时，有可能会出现错误提示

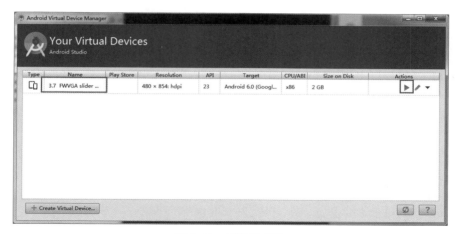

图 3-43　启动模拟器

图 3-44　项目运行效果图

"Intel HAXM is required to run this AVD，VT-x is disabled in BIOS"，这里给出解决方式。

① 在 SDK Manager 中检查是否安装 Intel x86 Emulator Accelerator（HAXM installer），如果没有安装，则在 SDK Manager 中下载安装它。

② 在 SDK 目录（一般为 C:\Users\Administrator\AppData\Local\Android\sdk）下

沿着 extras\intel\Hardware_Accelerated_Execution_Manager 目录找到 intelhaxm-android.exe 这个文件，安装并运行。

③ 在安装过程中，你可能会遇到这样的错误提示 Intel virtualization technology（vt，vt-x）is not enabled。不用担心，这时你只需要进入 BIOS 界面，在 Configuration 选项中找到 Intel Virtualization Technology 并将其设置成 Enable 即可。

④ 重新启动 Android Studio，然后再次启动 AVD 即可。如果成功，可能要花几分钟的时间去显示虚拟器的窗口。

（2）很多人运行 Android Studio 时，系统会提醒 JDK 或 Android SDK 不存在，这时需要通过全局的 Project Structure 页面进行重新设置。选择 Configure→Project Defaults→Project Structure，进入全局的 Project Structure 页面。在此页面下设置 JDK 或 Android SDK 目录即可，如图 3-45 所示。

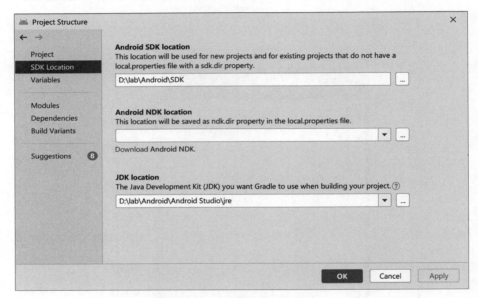

图 3-45　配置 SDK 和 JDK 路径

3.3　本地数据版 App 案例

"豹考通"App 分为本地版和网络版，两个版本的主要区别是本地版显示的主要数据来源于 App 本身，网络版数据主要来源于网络服务器；本地版显示的数据比网络版要少，但是显示界面基本类似。下面介绍如何运行"豹考通"本地版。

（1）打开 Android Studio 和虚拟模拟器，如图 3-46 所示。
（2）从提供的资源网站下载本地版客户端代码包并进行解压。
（3）选择 File 菜单下的 Open 选项（如图 3-47 所示），弹出选择对话框。
（4）在弹出的列表中选择客户端所在位置，单击 OK 按钮导入项目，如图 3-48 所示。
（5）项目部署在模拟器中，运行效果如图 3-49～图 3-54 所示。

第 3 章　App 应用体验

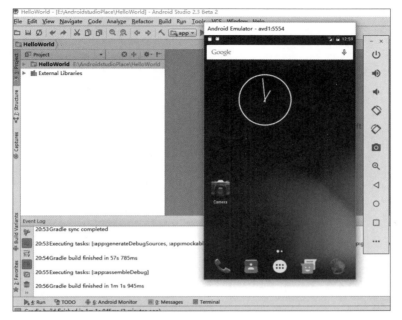

图 3-46　打开 Android Studio 和虚拟模拟器

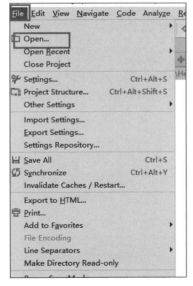

图 3-47　选择 Open 选项

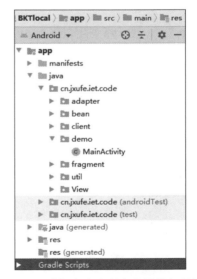

图 3-48　选择客户端所在目录导入项目

图 3-49　查询学校历年录取线

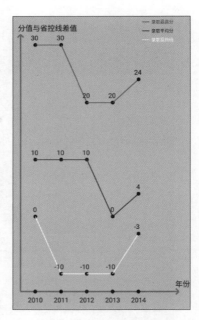

图 3-50　学校历年录取线趋势图

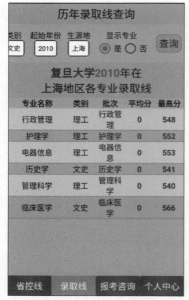

图 3-51　学校各专业录取线

图 3-52　查询各地区省控线信息

第 3 章　App 应用体验

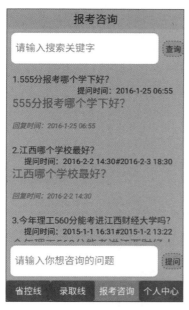

图 3-53　查询和提交报考咨询

图 3-54　个人基本信息编辑

3.4　网络数据版 App 案例

3.4.1　设计网络服务器

"豹考通"教学版项目包括服务器端、数据库和移动客户端三部分,体验客户端之前需要在本地安装服务器端和数据库。为了让初学者迅速体验 App 的功能效果,本书把服务器端和数据库放在 JTM 集成环境中运行。

JTM 是绿色、免费的 JDK＋Tomcat＋MySQL 环境集成工具。通过 JTM,用户无须对 JDK、Tomcat、MySQL 进行任何安装和配置即可迅速搭建支持 JSP＋MySQL 的服务器运行环境。下面介绍如何使用"豹考通"教学版 JTM。

(1) 从提供的资源网站下载 JTM 压缩包并解压到不带有空格的路径上(见图 3-55 和图 3-56)。

图 3-55　"豹考通"JTM 压缩包　　　　图 3-56　"豹考通"JTM 解压文件

(2) 双击启动 JTM.exe 文件,启动后的效果如图 3-57 所示。此时,Tomcat 和 MySQL 前面的图标都是灰色的,表示本机的这两个服务暂未启动,为正常现象。如果这

45

两个图标为绿色,则表示本机已启动相关服务或者端口被占用,需要停止相应的服务或进程。

（3）单击"启动（调试模式）"按钮,将会弹出两个命令行窗口,分别表示 Tomcat 的启动和 MySQL 的启动,此时 Tomcat 和 MySQL 前面的图标将会转为绿色,如图 3-58 所示。同时还会启动一个浏览器显示 Tomcat 的首页。

图 3-57　JTM 未启动时正常的状态

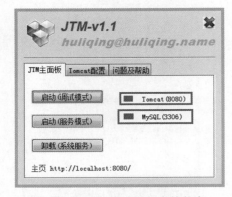

图 3-58　JTM 启动后正常的状态

（4）在浏览器的地址栏中输入 http://localhost:8080/bkt。此时,将会出现本地服务器测试成功页面,如图 3-59 所示。在使用过程中要始终保持这两个窗口处于开启状态,否则将无法访问和获取相关数据。

图 3-59　本地服务器搭建成功的测试页面

3.4.2 Android 功能实现

（1）从提供的资源网站下载网络版客户端代码包并进行解压。

（2）导入网络版客户端项目，如图 3-60 所示。

图 3-60　选择客户端所在目录导入项目

（3）导入项目后，打开 Global 类，找到 URL 语句，修改 IP 为计算机本地 IP，然后运行项目，如图 3-61 所示（提示：导入项目时，如果中文字符显示乱码，可以将 UTF-8 编码改成 GBK 编码）。

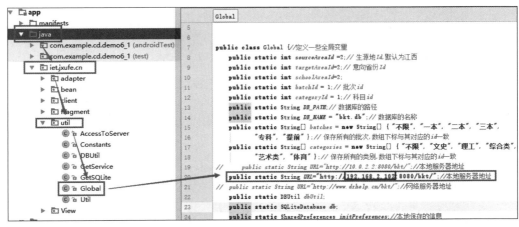

图 3-61　修改本地 IP 地址

3.4.3 客户端运行效果

图 3-62～图 3-67 为客户端运行效果图。

图 3-62　查询学校历年录取线

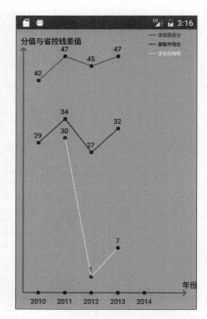

图 3-63　学校历年录取线趋势图

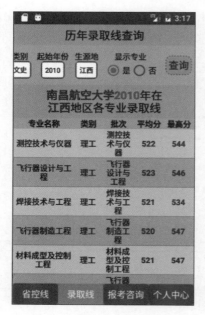

图 3-64　学校各专业录取线

图 3-65　查询各地区省控线信息

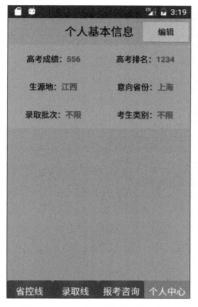
图 3-66　查询和提交报考咨询

图 3-67　个人基本信息编辑

3.5　本章小结

"豹考通"教学版分为本地版和网络版。本章介绍了如何搭建 Java 开发环境，下载 Android Studio 并安装使用。先是通过 Android Studio 导入本地版，运行体验"豹考通"App 项目。然后是通过 JTM 集成软件启动"豹考通"服务器端和数据库，再通过 Android Studio 导入网络版"豹考通"教学版，修改 IP 地址并运行体验。在后面的章节中，将逐步实现本地版，然后学习搭建数据库和服务器实现网络版"豹考通"。

3.6　课后练习

1. 新建 Android 项目 HelloWorld 并描述程序的执行过程。
2. 从应用商店下载并使用"豹考通"真实版 App。
3. 移动 App 开发流程一般分为哪几个部分？试简要说明之。
4. 描述 Android 程序下各个文件夹的作用。
5. 请设计并实现图 3-68 所示的界面效果。

界面要求：
- 页面背景颜色为♯aabbcc。最上面是一个 TextView 控件，用于显示标题信息，在标题文字的左边和右边各有一个图标，图标为应用图标，标题文字与图片居中显示。标题信息与顶部的边距为 10dp，标题文字为"欢迎进入全栈工程师实战案例教程学习"，标题文字大小为 18sp。

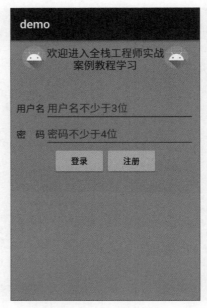

图 3-68　界面效果

- 中间部分包含两个文本显示框(TextView)、两个文本编辑框(EditText)和两个按钮(Button)。文本显示框的内容分别为"用户名"和"密码",文字大小为 16sp。第一个文本编辑框的提示信息为"用户名不少于 3 位",第二个文本编辑框是密码框,提示信息为"密码不少于 4 位"。两个按钮的内容分别为"登录"和"注册",水平居中显示。

第 4 章

Android 客户端设计和实现

4.1 本章简介

本章主要介绍如何实现本地版"豹考通"。"豹考通"分为"省控线查询""录取线查询""报考咨询""个人中心"四个模块。其中省控线查询和录取线查询比较类似,只选取一个重点讲解。在录取线查询中包含趋势图绘制,在此单独作为一节介绍其实现。四个模块需要实现底部导航栏灵活切换,本章单独进行实现。因此,本章划分为"查询界面"模块设计和实现、"报考咨询"模块设计和实现、"个人基本信息"模块设计和实现、多页面切换效果设计和实现以及绘制趋势图五部分讲解。在顺序方面,为了让知识点由浅入深,先实现四个页面,再把四个页面用底部导航栏组合在一起变成一个项目,最后实现录取线查询页面趋势图,如图 4-1 所示。

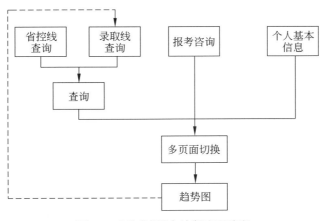

图 4-1 "豹考通"本地版实现流程

4.2 "查询界面"模块设计与实现

查询界面中包括省控线查询和录取线查询,本节以省控线查询作为代表进行讲解。在"省控线查询"模块中,先是实现模块整体界面布局,然后实现 4 个 Spinner 下拉列表。接下来对 4 个列表监听,实现省控线标题变化,最后用 ListView 控件静态实现省控线列表,如图 4-2 所示。

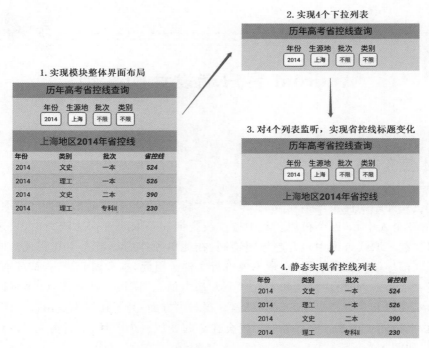

图 4-2 "省控线查询"模块实现的页面流程

4.2.1 用 Spinner 实现下拉列表选项

4 个下拉列表采用 Spinner 控件实现,如图 4-3 所示,每个 Spinner 列表需要自己的布局、数据源和适配器。在 4 个 Spinner 列表的布局中,使用了 3 种布局,其中年份和生源地的布局一样,因此创建 3 个布局文件即可。3 种布局文件比较相似,只是颜色不一样,为节省代码,使用 style 样式表进行实现。3 个布局文件名分别是 spinner_item_batch.xml、spinner_item_category.xml 和 spinner_item_ year.xml。

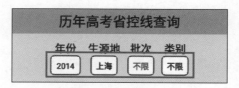

图 4-3 省控线查询的 Spinner 列表

首先,在样式表 styles.xml 文件中添加 spinner_itemStyle 样式,代码如程序清单 4-1 所示。

程序清单 4-1:Code040201\app\src\main\res\values\styles.xml

```
1    <style name="spinner_itemStyle">
2    <item name="android:layout_width">match_parent</item>
```

```
3    <item name="android:layout_height">wrap_content</item>
4    <item name="android:padding">5dp</item>
5    <item name="android:textSize">12sp</item>
6    <item name="android:gravity">center</item>
7    <item name="android:background">#ffffff</item>
8    <item name="android:textColor">#000000</item>
9  </style>
```

创建 spinner_item_batch.xml 文件，添加代码如程序清单 4-2 所示。

程序清单 4-2：Code040201\app\src\main\res\layout\spinner_item_batch.xml

```
1  <TextView xmlns:android="http://schemas.android.com/apk/res/android"
2      android:textColor="#ff0000"
3      style="@style/spinner_itemStyle">
4  </TextView>
```

创建 spinner_item_category.xml 文件，添加代码如程序清单 4-3 所示。

程序清单 4-3：Code040201\app\src\main\res\layout\spinner_item_category.xml

```
1  <TextView xmlns:android="http://schemas.android.com/apk/res/android"
2      android:textColor="#0000ff"
3      style="@style/spinner_itemStyle">
4  </TextView>
```

创建 spinner_item_year.xml 文件，添加代码如程序清单 4-4 所示。

程序清单 4-4：Code040201\app\src\main\res\layout\spinner_item_year.xml

```
1  <TextView xmlns:android="http://schemas.android.com/apk/res/android"
2      style="@style/spinner_itemStyle">
3  </TextView>
```

实现 Spinner 列表布局文件后，还需要在 Java 类文件中创建数据源和适配器。打开主类文件资源代码 ControlLineActivity（新建项目默认为 MainActivity），添加常量和实例化控件，如程序清单 4-5 所示。

程序清单 4-5：Code040201\app\src\main\java\cn\jxufe\iet\code040201\ControlLineActivity.java

```
1  private Spinner yearSpinner, provinceSpinner, categorySpinner,batchSpinner;
                                                              //下拉列表控件
2  private String[] years = new String[] { "2014", "2013", "2012", "2011","2010" };
                                                              //年份的集合
3  private List<String> provinces = new ArrayList<String>();  //省份的集合
4  private String [] batchs=new String[]{"不限","一本","二本","三本","专科",
   "提前"};
5  private String [] categorys=new String[]{"不限","文史","理工","综合类","艺术
```

类","体育"};
```
6     private int sourceAreaId,yearId,categoryId,batchId;
7     private TextView controlLineTitle;              //提示标题
8     private List<ControlLine> controlLineList;       //分数线的集合
9     private ListView controlLineListView;            //显示分数线的列表
```

在 ControlLineActivity 主类文件的 onCreate 方法中，关联控件获取 ID，代码如程序清单 4-6 所示。

程序清单 4-6：Code040201\app\src\main\java\cn\jxufe\iet\code040201\ControlLineActivity.java

```
1    yearSpinner = (Spinner) findViewById(R.id.yearSpinner);     //年份的下拉列表
2    provinceSpinner = (Spinner)findViewById(R.id.areaSpinner);  //省份的下拉列表
3    categorySpinner = (Spinner)findViewById(R.id.categorySpinner);
4    batchSpinner = (Spinner)findViewById(R.id.batchSpinner);
5    controlLineTitle = (TextView)findViewById(R.id.controlLineTitle);
                                                                  //显示结果标题
6    controlLineListView = (ListView)findViewById(R.id.controlLineList);
                                                                  //显示查询结果
```

由于生源地数据源在整个客户端项目中是从 SQLite 数据库获取，现在新建 GetSQLite 类，定义临时数据集合以代替 SQLite 生源地数据，代码如程序清单 4-7 所示。

程序清单 4-7：Code040201\app\src\main\java\cn\jxufe\iet\code040201\GetSQLite.java

```
1    public class GetSQLite {
2        /*把省份数据封装成集合,为后面调用数据库数据做准备*/
3        public List<String> setProvince(){
4            List<String> provinces=new ArrayList<>();
5            provinces.add("上海");
6            provinces.add("浙江");
7            provinces.add("江西");
8            provinces.add("北京");
9            return provinces;
10       }
11   }
```

在 ControlLineActivity 主类文件的 onCreate 方法中，实例化 GetSQLite 对象获取本地数据临时生源地集合。然后，定义 4 个 Spinner 列表适配器并将数据源和布局关联，这样就实现了 4 个 Spinner 列表，代码如程序清单 4-8 所示。

程序清单 4-8：Code040201\app\src\main\java\cn\jxufe\iet\code040201\ControlLineActivity.java

```
1    GetSQLite getSQLite=new GetSQLite();
2    provinces=getSQLite.setProvince();                    //获取省份集合
3
```

```
4   ArrayAdapter<String> provincesAdapter = new ArrayAdapter<String>(
5           this, R.layout.spinner_item_year, provinces);
6   provinceSpinner.setAdapter(provincesAdapter);
7   provinceSpinner.setSelection(sourceAreaId);        //设置默认省份显示项
8
9   ArrayAdapter<String> yearAdapter = new ArrayAdapter<String>(
10          this, R.layout.spinner_item_year, years);
11  yearSpinner.setAdapter(yearAdapter);
12
13  ArrayAdapter<String> categoryAdapter = new ArrayAdapter<String>(
14          this, R.layout.spinner_item_category,categorys);
15  categorySpinner.setAdapter(categoryAdapter);
16  categorySpinner.setSelection(categoryId);          //设置默认科类显示项
17
18  ArrayAdapter<String> batchAdapter = new ArrayAdapter<String>(
19          this, R.layout.spinner_item_batch, batchs);
20  batchSpinner.setAdapter(batchAdapter);
21  batchSpinner.setSelection(batchId);                //设置默认批次显示项
```

4.2.2 事件监听器

4.2.1 节实现了 4 个 Spinner 列表,但还不能实现在选择列表后使省控线标题相应切换的效果(如图 4-4 所示)。为了使省控线随着选择列表的不同而相应变化,可采用内部类事件监听方法,对 4 个列表进行监听。首先在主类 ControlLineActivity 中定义 MyItemSelectedListener 类并实现 OnItemSelectedListener 类的接口,在执行完 onItemSelected 方法后显示省控线标题信息,代码如程序清单 4-9 所示。

图 4-4　省控线查询事件监听

程序清单 4-9:Code040201\app\src\main\java\cn\jxufe\iet\code040201\ControlLineActivity.java

```
1   private class MyItemSelectedListener implements AdapterView
        .OnItemSelectedListener {
2       @Override
3       public void onItemSelected(AdapterView<?> parent, View view,
4                                  int position, long id) {
5           switch (parent.getId()) {
```

```
6              case R.id.areaSpinner:         //如果是选择省份列表
7                  sourceAreaId = position;
8                  break;
9              case R.id.yearSpinner:         //如果是选择年份列表
10                 yearId = position;
11                 break;
12             case R.id.categorySpinner:     //如果是选择科类列表
13                 categoryId = position;
14                 break;
15             case R.id.batchSpinner:        //如果是选择批次列表
16                 batchId = position;
17                 break;
18             default:
19                 break;
20         }
21         /*设置标题显示*/
22         controlLineTitle.setText(Html.fromHtml("<font color=red><b>"
23                 + provinces.get(sourceAreaId)
24                 + "</b></font>地区<font color=blue><b>" + years[yearId]
25                 + "</b></font>年省控线"));
26     }
27     @Override
28     public void onNothingSelected(AdapterView<?> parent) {
29     }
30 }
```

定义完 MyItemSelectedListener 事件监听内部类后，需要对 4 个 Spinner 控件监听才能达到监听的效果，代码如程序清单 4-10 所示。

程序清单 4-10：Code040201\app\src\main\java\cn\jxufe\iet\code040201\ControlLineActivity.java

```
1  MyItemSelectedListener itemSelectedListener = new MyItemSelectedListener();
2  yearSpinner.setOnItemSelectedListener(itemSelectedListener);
3  provinceSpinner.setOnItemSelectedListener(itemSelectedListener);
4  categorySpinner.setOnItemSelectedListener(itemSelectedListener);
5  batchSpinner.setOnItemSelectedListener(itemSelectedListener);
```

4.2.3 ListView 列表

省控线 ListView 列表需要分步进行实现，需要实现列表布局、数据源和列表适配器，如图 4-5 所示。

在 layout 文件夹下新建 list_item_control_line.xml 列表布局，代码如程序清单 4-11 所示。

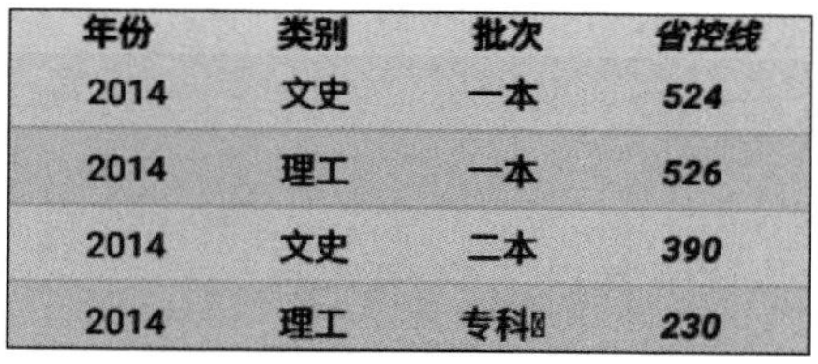

年份	类别	批次	省控线
2014	文史	一本	524
2014	理工	一本	526
2014	文史	二本	390
2014	理工	专科	230

图 4-5　省控线查询列表

程序清单 4-11：Code040202\app\src\main\res\layout\list_item_control_line.xml

```
1   <LinearLayout xmlns:android="http://schemas.android.com/apk/res/android"
2       android:layout_width="match_parent"
3       android:layout_height="wrap_content"
4       android:layout_marginLeft="30dp"
5       android:gravity="center_horizontal"
6       android:orientation="horizontal"
7       android:background="#33aabbcc"
8       android:padding="5dp">
9
10  <TextView
11      android:id="@+id/year"
12      android:layout_width="0dp"
13      android:layout_height="wrap_content"
14      android:layout_marginLeft="10dp"
15      android:layout_marginRight="10dp"
16      android:layout_weight="1"
17      android:textColor="#000000"
18      android:gravity="center" />
19
20  <TextView
21      android:id="@+id/category"
22      android:layout_width="0dp"
23      android:layout_height="wrap_content"
24      android:layout_marginRight="10dp"
25      android:layout_weight="1"
26      android:textColor="#000000"
27      android:gravity="center" />
28
29  <TextView
30      android:id="@+id/batch"
31      android:layout_width="0dp"
32      android:layout_height="wrap_content"
33      android:layout_marginRight="10dp"
34      android:layout_weight="1"
```

```
35              android:textColor="#000000"
36              android:gravity="center" />
37
38      <TextView
39              android:id="@+id/controlLine"
40              android:layout_width="0dp"
41              android:layout_height="wrap_content"
42              android:layout_marginRight="10dp"
43              android:layout_weight="1"
44              android:gravity="center"
45              android:textColor="#0000ff"
46              android:textStyle="bold|italic"/>
47  </LinearLayout>
```

在"豹考通"整体项目中,列表数据源是从服务端获取数据,现在使用临时集合数据代替从服务端获取的数据,新建 GetService 类添加集合数据,在集合中使用 ControlLine 实体类封装每条数据。

ControlLine 类代码如程序清单 4-12 所示。

程序清单 4-12: Code040202\app\src\main\java\cn\jxufe\ief\code040202\bean\ControlLine.java

```
1   public class ControlLine implements Serializable {      //省控线封装类
2       private static final long serialVersionUID = 7397428155350981454L;
3       private int controlLine;                            //省控线
4       private String areaName;                            //省份名
5       private String batchName;                           //批次名
6       private String categoryName;                        //类别名
7       private int controlYear;                            //年份
8       public int getControlLine() {
9           return controlLine;
10      }
11      public void setControlLine(int controlLine) {
12          this.controlLine = controlLine;
13      }
14      public String getAreaName() {
15          return areaName;
16      }
17      public void setAreaName(String areaName) {
18          this.areaName = areaName;
19      }
20      public String getBatchName() {
21          return batchName;
22      }
23      public void setBatchName(String batchName) {
24          this.batchName = batchName;
```

```
25      }
26      public String getCategoryName() {
27          return categoryName;
28      }
29      public void setCategoryName(String categoryName) {
30          this.categoryName = categoryName;
31      }
32      public int getControlYear() {
33          return controlYear;
34      }
35      public void setControlYear(int controlYear) {
36          this.controlYear = controlYear;
37      }
38      @Override
39      public String toString() {
40          return "ControlLine [cotrolLine=" + controlLine + ", areaName="
41                  + areaName + ", batchName=" + batchName + ", categoryName="
42                  + categoryName + ", controlYear=" + controlYear + "]";
43      }
44  }
```

GetService 类代码如程序清单 4-13 所示。

程序清单 4-13：Code040202\app\src\main\java\cn\jxufe\iet\code040202\util\GetService.java

```
1   public class GetService {
2       private int[] yearString=new int[]{2014,2014,2014,2014};
3       private String[] categroyString=new String[]{"文史","理工","文史","理工"};
4       private String[] batchString=new String[]{"一本","一本","二本","专科"};
5       private int[] scoreString=new int[]{524,526,390,230};
6       private String[] areaNameString=new String[]{"江西","江西","江西","江西"};
7       /*设置省控线数据集合*/
8       public List<ControlLine> setControlLineList(){
9           List<ControlLine> list=new ArrayList<ControlLine>();
10          for (int i=0;i<4;i++){
11              ControlLine controlLine=new ControlLine();
12              controlLine.setControlYear(yearString[i]);
13              controlLine.setCategoryName(categroyString[i]);
14              controlLine.setBatchName(batchString[i]);
15              controlLine.setControlLine(scoreString[i]);
16              controlLine.setAreaName(areaNameString[i]);
17              list.add(controlLine);
18          }
19          return list;
20      }
```

```
21   }
```

ListView 列表采用自定义列表,需要实现自定义 Adapter 类并继承 BaseAdapter 类。新建 ControlLineAdapter 类,实现代码如程序清单 4-14 所示。

程序清单 4-14:Code040202\app\src\main\java\cn\jxufe\iet\code040202\adapter\ControlLineAdapter.java

```java
1   public class ControlLineAdapter extends BaseAdapter {
2       private Context context;
3       private List<ControlLine> controlLineList;
4       public ControlLineAdapter
5           (Context context, List<ControlLine> controlLineList) {
6           //TODO Auto-generated constructor stub
7           this.context=context;
8           this.controlLineList=controlLineList;
9       }
10      @Override
11      public int getCount() {
12          //TODO Auto-generated method stub
13          return controlLineList.size();
14          //返回集合数据数量
15      }
16      @Override
17      public Object getItem(int position) {
18          //TODO Auto-generated method stub
19          return controlLineList;                      //返回子项
20      }
21      @Override
22      public long getItemId(int position) {
23          //TODO Auto-generated method stub
24          return position;                             //返回子项 Id
25      }
26      @Override
27      public View getView(int position, View convertView, ViewGroup parent) {
28          //TODO Auto-generated method stub
29          ViewHolder holder=new ViewHolder();
30          if (convertView == null) {
31              convertView = LayoutInflater.from(this.context)
        .inflate(R.layout.list_item_control_line, null, false);
32              if (position%2==0) {
33                  //设置奇偶项背景颜色
34                  convertView.setBackgroundColor(Color.argb(200, 220, 230, 240));
35              }else{
36                  convertView.setBackgroundColor(Color.argb(51, 170, 187, 204));
```

```
37        }
38            /**得到各个控件的对象*/
39            holder.year=(TextView)convertView.findViewById(R.id.year);
40            holder.category=(TextView)convertView.findViewById(R.id.category);
41            holder.batch=(TextView)convertView.findViewById(R.id.batch);
42            holder.controlLine=(TextView)convertView.findViewById(R.id.controlLine);
43            convertView.setTag(holder);          //绑定 ViewHolder 对象
44        }else {
45            holder = (ViewHolder)convertView.getTag();
46            //取出 ViewHolder 对象
47        }
48        holder.year.setText(controlLineList.get(position)
   .getControlYear()+"");
49        holder.category.setText(controlLineList.get(position)
   .getCategoryName());
50        holder.batch.setText(controlLineList.get(position).getBatchName());
51        holder.controlLine.setText(controlLineList.get(position).
   getControlLine()+"");
52        return convertView;
53    }
54    /**存放控件*/
55    public final class ViewHolder{
56        public TextView year;
57        public TextView category;
58        public TextView batch;
59        public TextView controlLine ;
60    }
61 }
```

在主类 ControlLineActivity 中实例化 GetService 对象获取省控线数据集合并实例化 ControlLineAdapter 对象,传入数据源,关联 ListView 控件显示列表,具体代码如程序清单 4-15 所示。

程序清单 4-15:Code040202\app\src\main\java\cn\jxufe\iet\code040202\ControlLineActivity.java

```
1 /* 显示省控线列表 */
2 GetService getService=new GetService();
3 controlLineList=getService.setControlLineList();
                                   //为 controlLineList 集合添加集合内容
4 ControlLineAdapter controlLineAdapter=new ControlLineAdapter
5        (this,controlLineList);
6 controlLineListView.setAdapter(controlLineAdapter);
7 //关联 ListView 控件显示列表
```

4.3 "报考咨询"模块设计与实现

在"报考咨询"模块中,先是实现模块界面整体布局,然后实现问题列表,对问题列表长按进行事件监听,长按能够弹出问题回复对话框,如图 4-6 所示。

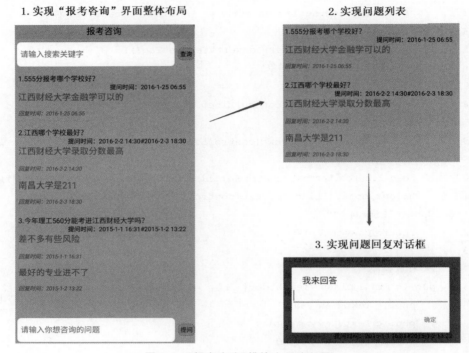

图 4-6 "报考咨询"模块实现流程图

4.3.1 界面设计

"报考咨询"模块整体采用线性布局,布局内部结合 TextView(文本显示框)、EditText(编辑框)、Button(按钮)、ListView(列表)等组件组合实现,具体参见图 4-7。

打开 strings.xml 文件,在 resources 标签下定义资源变量,代码如程序清单 4-16 所示。

程序清单 4-16:Code040301\app\ src\main\res\values\strings.xml

```
1    <resources>
2    <string name="app_name">Code0501</string>
3    <string name="consult">报考咨询</string>
4    <string name="submitQuestion">提问</string>
5    <string name="inputHint">请输入你想咨询的问题</string>
6    <string name="keywordHint">请输入搜索关键字</string>
```

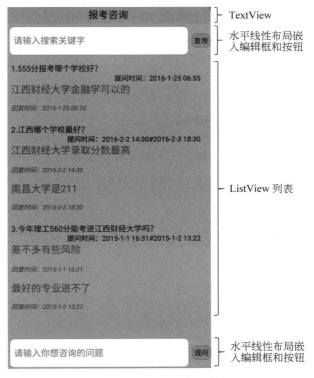

图 4-7 "报考咨询"模块整体布局

```
7    <string name="save">保存</string>
8    <string name="selfInfoTitle">个人基本信息</string>
9    <string name="score">高考成绩</string>
10   <string name="inputScore">输入高考成绩</string>
11   <string name="rank">高考排名</string>
12   <string name="targetArea">意向省份</string>
13   <string name="recruitBatch">录取批次</string>
14   <string name="studentCategory">考试类别</string>
15   <string name="search">查询</string>
16   </resources>
```

在"报考咨询"模块中有两个编辑框,编辑框的边框使用了背景图片,这张背景图片为自定义图片,采用 selector 标签根据状态显示图片,分别为 edit_test_pressed.png 与 edit_test_normal.png。新建 edit_text_bg.xml 图片,具体代码如程序清单 4-17 所示。

程序清单 4-17:Code040301\app\src\main\res\drawable\edit_text_bg.xml

```
1    <selector
2        xmlns:android="http://schemas.android.com/apk/res/android">
3    <item android:state_focused="true"
4         android:drawable="@drawable/edit_text_pressed" />
```

```
5    <item android:state_focused="false"
6          android:state_pressed="true"
7          android:drawable="@drawable/edit_text_pressed" />
8    <item android:state_focused="false"
9          android:drawable="@drawable/edit_text_normal" />
10   </selector>
```

模块中的"查询"和"提交"按钮使用了背景图片 title_btn.xml,此背景图片为使用 selector 标签根据状态显示两张不同背景图片的自定义图片。

"报考咨询"模块整体布局代码如程序清单 4-18 所示。

程序清单 4-18:Code040301\app\src\main\res\layout\activity_main.xml

```
1    <LinearLayout xmlns:android="http://schemas.android.com/apk/res/android"
2        xmlns:tools="http://schemas.android.com/tools"
3        android:layout_width="match_parent"
4        android:layout_height="match_parent"
5        android:background="#66ddccbb"
6        android:orientation="vertical"
7        tools:context=".ConsultActivity">
8        <TextView
9            android:layout_width="match_parent"
10           android:layout_height="wrap_content"
11           android:background="#ccbbaa"
12           android:gravity="center"
13           android:padding="5dp"
14           android:text="@string/consult"
15           android:textColor="#000000"
16           android:textSize="20sp" />
17       <LinearLayout
18           android:layout_width="match_parent"
19           android:layout_height="wrap_content"
20           android:gravity="center_vertical"
21           android:orientation="horizontal">
22           <EditText
23               android:id="@+id/keyword"
24               android:layout_width="0dp"
25               android:layout_height="wrap_content"
26               android:layout_margin="5dp"
27               android:layout_weight="1"
28               android:background="@drawable/edit_text_bg"
29               android:textColor="#000000"
30               android:hint="@string/keywordHint" />
31           <Button
32               android:id="@+id/search"
```

```xml
33              android:layout_width="wrap_content"
34              android:layout_height="wrap_content"
35              android:background="@drawable/title_btn"
36              android:gravity="center"
37              android:minHeight="0dp"
38              android:minWidth="0dp"
39              android:padding="5dp"
40              android:text="@string/search" />
41      </LinearLayout>
42      <ListView
43              android:id="@+id/questionList"
44              android:layout_width="match_parent"
45              android:layout_height="0dp"
46              android:layout_weight="1"
47              android:divider="#00ffffff"
48              android:dividerHeight="5dp" />
49      <LinearLayout
50              android:layout_width="match_parent"
51              android:layout_height="wrap_content"
52              android:gravity="center_vertical"
53              android:orientation="horizontal">
54      <EditText
55              android:id="@+id/content"
56              android:layout_width="0dp"
57              android:layout_height="wrap_content"
58              android:layout_margin="5dp"
59              android:layout_weight="1"
60              android:background="@drawable/edit_text_bg"
61              android:textColor="#000000"
62              android:hint="@string/inputHint" />
63      <Button
64              android:id="@+id/submit"
65              android:layout_width="wrap_content"
66              android:layout_height="wrap_content"
67              android:background="@drawable/title_btn"
68              android:gravity="center"
69              android:minHeight="0dp"
70              android:minWidth="0dp"
71              android:padding="5dp"
72              android:text="@string/submitQuestion" />
73      </LinearLayout>
74      </LinearLayout>
```

4.3.2 问题列表的实现

"报考咨询"模块的 ListView 列表需要分步进行实现，需要实现列表布局、数据源和列表适配器，如图 4-8 所示。

图 4-8 "报考咨询"模块的问题列表

在 layout 文件夹下新建 list_item_question.xml 列表布局，代码如程序清单 4-19 所示。

程序清单 4-19：Code040301\app\src\main\res\layout\list_item_question.xml

```
1    <LinearLayout
2        xmlns:android="http://schemas.android.com/apk/res/android"
3        xmlns:tools="http://schemas.android.com/tools"
4        android:layout_width="match_parent"
5        android:layout_height="wrap_content"
6        android:padding="10dp"
7        android:orientation="vertical">
8        <TextView
9            android:id="@+id/questionContent"
10           android:layout_width="match_parent"
11           android:layout_height="wrap_content"
12           android:textSize="16sp"
13           android:textColor="#0000ff"/>
14       <TextView
15           android:id="@+id/questionTime"
16           android:layout_width="match_parent"
17           android:layout_height="wrap_content"
18           android:paddingRight="10dp"
19           android:textColor="#000000"
20           android:gravity="right"/>
21       <TextView
22           android:id="@+id/answer"
```

```
23              android:layout_width="match_parent"
24              android:layout_height="wrap_content"
25              android:textSize="16sp"
26              android:textColor="#ff0000"/>
27  </LinearLayout>
```

在"豹考通"整体项目中，列表数据源是从服务端获取数据，现在使用临时集合数据代替从服务端获取的数据，新建 GetService 类添加集合数据，在集合中使用 Question 实例类封装每条数据。

Question 类代码如程序清单 4-20 所示。

程序清单 4-20：Code040301\app\src\main\java\cn\jxufe\iet\code040301\bean\Question.java

```
1   public class Question implements Serializable {        ////报考咨询中的问题
2       private static final long serialVersionUID = 1L;
3       public static final String QUERY="query";
4       public static final String INSERT="insert";
5       public static final String UPDATE="update";
6       private int questionId;                             //问题 id
7       private String questionContent;                     //问题内容
8       private String questionTime;                        //提问时间
9       private String answerContent;                       //问题回复
10      private String answerTime;                          //回复时间
11      private String questionStatus;                      //问题状态:待回答、已回答
12      public int getQuestionId() {
13          return questionId;
14      }
15      public void setQuestionId(int questionId) {
16          this.questionId = questionId;
17      }
18      public String getQuestionContent() {
19          return questionContent;
20      }
21      public void setQuestionContent(String questionContent) {
22          this.questionContent = questionContent;
23      }
24      public String getQuestionTime() {
25          return questionTime;
26      }
27      public void setQuestionTime(String questionTime) {
28          this.questionTime = questionTime;
29      }
30      public String getAnswerContent() {
31          return answerContent;
32      }
```

```java
33      public void setAnswerContent(String answerContent) {
34          this.answerContent = answerContent;
35      }
36      public String getAnswerTime() {
37          return answerTime;
38      }
39      public void setAnswerTime(String answerTime) {
40          this.answerTime = answerTime;
41      }
42      public String getQuestionStatus() {
43          return questionStatus;
44      }
45      public void setQuestionStatus(String questionStatus) {
46          this.questionStatus = questionStatus;
47      }
48      @Override
49      public String toString() {
50          return "Question [questionId=" + questionId + ", questionContent="
51              + questionContent + ", questionTime=" + questionTime
52              + ", answerContent=" + answerContent + ", answerTime="
53              + answerTime + ", questionStatus=" + questionStatus + "]";
54      }
55  }
```

GetService 类代码如程序清单 4-21 所示。

程序清单 4-21：Code040301\app\src\main\java\cn\jxufe\iet\code040301\util\GetService.java

```java
1   public class GetService {
2       //问题咨询数组
3       private String[] questionContent=new String []{"555分报考哪个学校好？",
4   "江西哪个学校最好？","今年理工560分能考进江西财经大学吗？"};  //问题数组
5       private String [] questionTime=new String[]
6           {"2016-1-25 03:55","2016-2-2 13:50","2015-1-1 14:20"};
                                                                    //提问时间数组
7       private String[] answerContent=new String[]{"江西财经大学金融学可以的",
8   "江西财经大学录取分数最高#南昌大学是211","差不多有些风险#最好的专业进不了"};
                                                                    //回答问题数组
9       private String[] answerTime=new String[]
10          {"2016-1-25 06:55","2016-2-2 14:30#2016-2-3 18:30",
11  "2015-1-1 16:31#2015-1-2 13:22"};                               //回复时间数组
12      /*把问题咨询数组数据转成集合数据*/
13      public List<Question> setQuestionList(){
14          List<Question> questionList=new ArrayList<>();
15
```

```
16          for(int i=0;i<questionContent.length;i++){
17              Question question=new Question();
18              question.setQuestionContent(questionContent[i]);
19              question.setQuestionTime(questionTime[i]);
20              question.setAnswerContent(answerContent[i]);
21              question.setAnswerTime(answerTime[i]);
22              questionList.add(question);
23          }
24          return questionList;
25      }
26  }
```

ListView 列表采用自定义列表,需要实现自定义 Adapter 类并继承 BaseAdapter 类。新建 ConsultAdapter 类,实现代码如程序清单 4-22 所示。

程序清单 4-22：Code040301\app\src\main\java\cn\jxufe\iet\code040301\adapter\ConsultAdapter.java

```
1   public class ConsultAdapter extends BaseAdapter {
2       private Context context;
3       private List<Question> questionList;              //专业录取线列表
4       public ConsultAdapter
5           (Context context,List<Question> questionList) {
6           //TODO Auto-generated constructor stub
7           this.context=context;
8           this.questionList=questionList;
9       }
10      @Override
11      public int getCount() {                           //获取列表数量
12          //TODO Auto-generated method stub
13          return questionList.size();
14      }
15      @Override
16      public Object getItem(int position) {             //获取列表子项
17          //TODO Auto-generated method stub
18          return questionList;
19      }
20      @Override
21      public long getItemId(int position) {             //获取列表项 id
22          //TODO Auto-generated method stub
23          return position;
24      }
25      @Override
26      public View getView(int position, View convertView, ViewGroup parent) {
27          //TODO Auto-generated method stub
28          ViewHolder holder=new ViewHolder();           //实例化 ViewHolder 对象
```

```java
29      if (convertView == null) {
30          //将布局转换成View对象
31          convertView = LayoutInflater.from(this.context)
32                  .inflate(R.layout.list_item_question, null, false);
33          //关联Id
34  holder.contentView=(TextView)convertView.findViewById(R.id.questionContent);
35          holder.timeView=(TextView)convertView.findViewById(R.id.questionTime);
36          holder.answerView=(TextView) convertView.findViewById(R.id.answer);
37          convertView.setTag(holder);
38      }else{
39          holder = (ViewHolder)convertView.getTag();  //取出ViewHolder对象
40      }
41      //设置对应显示对象
42      holder.contentView.setText(position + 1 + "."
43              + questionList.get(position).getQuestionContent());
                                                            //显示问题内容
44      holder.timeView.setText("提问时间:" + questionList.get(position).getAnswerTime());
45      //显示问题时间
46      if ("".equals(questionList.get(position).getAnswerTime())
47              || null == questionList.get(position).getAnswerTime()) {
48          holder.answerView.setText("暂无回复");
49      } else {
50          String[] answers = questionList.get(position).getQuestionContent().split("#");
51          //将多个回复分隔开
52          String[] answerTimes = questionList.get(position).getAnswerTime().split("#");
53          //将多个回复的时间分隔开
54          String answerResult = "";
55          for (int i = 0; i < answers.length; i++) {//显示回复内容和时间
56              answerResult += "<p><big>" + answers[i]
57                      + "</big></p><i><small>回复时间:" + answerTimes[i]
58                      + "</small></i>";
59          }
60          holder.answerView.setText(Html.fromHtml(answerResult));
61      }
62      return convertView;
63  }
64  /**存放控件*/
65  public final class ViewHolder{
66      public TextView contentView;
```

```
67          public TextView timeView;
68          public TextView answerView;
69      }
70  }
```

打开主类 ConsultActivity 定义常量信息,代码如程序清单 4-23 所示。

程序清单 4-23: Code040301\app\src\main\java\cn\jxufe\iet\code040301\ConsultActivity.java

```
1   private ListView questionListView;          //问题列表
2   private Button submit, search;              //提交问题按钮、查询按钮
3   private String questionString;              //问题字符串
4   private EditText contentView;               //提交内容编辑框
5   private EditText keywordView;               //搜索编辑框
6   private List<Question> questionList;
```

在主类 ConsultActivity 的 onCreate 方法中,获取组件 Id,代码如程序清单 4-24 所示。

程序清单 4-24: Code040301\app\src\main\java\cn\jxufe\iet\code040301\ConsultActivity.java

```
1   questionListView = (ListView) findViewById(R.id.questionList);
2   //根据 id 获取列表控件
3   contentView = (EditText) findViewById(R.id.content);
4   //获取编辑框按钮 id
5   submit = (Button) findViewById(R.id.submit);
6   //获取提交按钮 Id
7   search = (Button) findViewById(R.id.search);
8   //根据 id 找到控件
9   keywordView = (EditText) findViewById(R.id.keyword);
10  //获取查询编辑框 id
```

在主类 ConsultActivity 中实例化 GetService 对象获取省控线数据集合并实例化 ConsultAdapter 对象,传入数据源,关联 ListView 控件显示列表,具体代码如程序清单 4-25 所示。

程序清单 4-25: Code040301\app\src\main\java\cn\jxufe\iet\code040301\ConsultActivity.java

```
1   GetService getService=new GetService();
2   questionList=getService.setQuestionList();
3   ConsultAdapter consultAdapter=new ConsultAdapter
4           (ConsultActivity.this,questionList);
5   questionListView.setAdapter(consultAdapter);
```

4.3.3 问题回复对话框的实现

问题回复对话框是在长按列表子项后弹出的,如图 4-9 所示。因此,需要对列表进行

长按事件监听,对列表子项长按事件监听方法的具体实现可查看程序清单 4-26 所示的代码。

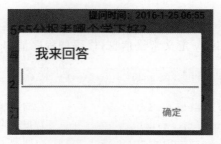

图 4-9　问题回复对话框

程序清单 4-26：Code040301\app\src\main\java\cn\jxufe\iet\code040301\ConsultActivity.java

```
1    questionListView.setOnItemLongClickListener(new AdapterView
     .OnItemLongClickListener() {
2        //长按某一项进行回复
3        @Override
4        public boolean onItemLongClick(AdapterView<?> parent,
5            View view, int position, long id) {
6            updateQuestion();                    //更新问题状态
7            return false;
8        }
9    });
```

在上面的代码中我们发现 updateQuestion()方法还没有实现,updateQuestion()是问题更新方法,在此方法中实现问题回复对话框,具体代码如程序清单 4-27 所示。

程序清单 4-27：Code040301\app\src\main\java\cn\jxufe\iet\code040301\ConsultActivity.java

```
1    public void updateQuestion() {              //更新问题状态
2        AlertDialog.Builder builder = new AlertDialog.Builder(ConsultActivity
     .this);                                      //实例化 Builder 对象
3        builder.setTitle("我来回答");
4        final EditText answerText = new EditText(ConsultActivity.this);
                                                  //实例化 EditText 对象
5        builder.setView(answerText);            //把编辑框以 View 的形式加入对话框
6        builder.setPositiveButton("确定", new DialogInterface.OnClickListener() {
                                                  //设置"确定"按钮
7            @Override
8            public void onClick(DialogInterface dialog, int which) {
9                //具体实现在第 7 章讲解
10           }
11       }
```

```
12          );
13          builder.create().show();              //创建对话框
14      }
```

4.4 "个人基本信息"模块设计和实现

4.4.1 界面设计

"个人基本信息"模块整体采用表格布局,布局内部结合 RelativeLayout(相对布局)、TextView(文本显示框)、Button(按钮)等组件实现,具体请参见图 4-10。

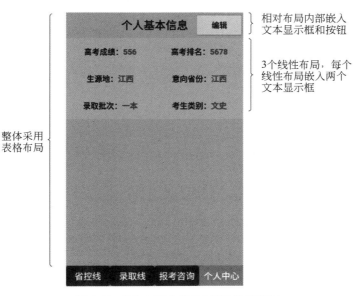

图 4-10 "个人基本信息"模块整体布局

打开 strings.xml 文件,在 resources 标签下定义资源变量,代码如程序清单 4-28 所示。

程序清单 4-28:Code040401\app\src\main\res\values\strings.xml

```
1   <resources>
2   <string name="selfInfoTitle">个人基本信息</string>
3   <string name="edit">编辑</string>
4   <string name="score">高考成绩</string>
5   <string name="inputScore">输入高考成绩</string>
6   <string name="rank">高考排名</string>
7   <string name="targetArea">意向省份</string>
8   <string name="recruitBatch">录取批次</string>
9   <string name="studentCategory">考试类别</string>
10  <string name="batch">批次</string>
11  <string name="category">类别</string>
```

```
12    <string name="sourceArea">生源地</string>
13  </resources>
```

"个人基本信息"模块整体布局代码如程序清单 4-29 所示。

程序清单 4-29：Code040401\app\src\main\res\layout\activity_main.xml

```
1   <TableLayout xmlns:android="http://schemas.android.com/apk/res/android"
2       xmlns:tools="http://schemas.android.com/tools"
3       android:layout_width="match_parent"
4       android:layout_height="wrap_content"
5       android:layout_gravity="center_horizontal"
6       android:background="#66ddccbb"
7       tools:context="cn.jxufe.iet.code0502.SelfInfoActivity">
8   <RelativeLayout
9       android:layout_width="match_parent"
10      android:layout_height="wrap_content"
11      android:background="#ccbbaa">
12  <TextView
13      android:layout_width="match_parent"
14      android:layout_height="wrap_content"
15      android:layout_centerInParent="true"
16      android:gravity="center"
17      android:padding="5dp"
18      android:singleLine="true"
19      android:text="@string/selfInfoTitle"
20      android:textColor="#000000"
21      android:textSize="20sp" />
22  <Button
23      android:id="@+id/edit"
24      android:layout_width="wrap_content"
25      android:layout_height="wrap_content"
26      android:layout_alignParentRight="true"
27      android:layout_centerVertical="true"
28      android:text="@string/edit" />
29  </RelativeLayout>
30  <LinearLayout
31      android:layout_width="match_parent"
32      android:layout_height="wrap_content"
33      android:layout_marginBottom="15dp"
34      android:layout_marginTop="15dp"
35      android:orientation="horizontal">
36  <!-- 高考分数 -->
37  <TextView
38      android:id="@+id/scoreText"
```

```xml
39            android:layout_width="0dp"
40            android:layout_height="wrap_content"
41            android:layout_weight="1"
42            android:gravity="center"
43            android:textColor="#000000" />
44    <!-- 高考排名 -->
45    <TextView
46            android:id="@+id/rankText"
47            android:layout_width="0dp"
48            android:layout_height="wrap_content"
49            android:layout_weight="1"
50            android:gravity="center"
51            android:textColor="#000000" />
52    </LinearLayout>
53    <LinearLayout
54            android:layout_width="match_parent"
55            android:layout_height="wrap_content"
56            android:layout_marginBottom="15dp"
57            android:layout_marginTop="15dp"
58            android:orientation="horizontal">
59    <!-- 生源地 -->
60    <TextView
61            android:id="@+id/sourceAreaText"
62            android:layout_width="0dp"
63            android:layout_height="wrap_content"
64            android:layout_weight="1"
65            android:gravity="center"
66            android:textColor="#000000" />
67    <!-- 意向省份 -->
68    <TextView
69            android:id="@+id/targetAreaText"
70            android:layout_width="0dp"
71            android:layout_height="wrap_content"
72            android:layout_weight="1"
73            android:gravity="center"
74            android:textColor="#000000" />
75    </LinearLayout>
76    <LinearLayout
77            android:layout_width="match_parent"
78            android:layout_height="wrap_content"
79            android:layout_marginBottom="15dp"
80            android:layout_marginTop="15dp"
81            android:orientation="horizontal">
82    <!-- 录取批次 -->
```

```
83  <TextView
84          android:id="@+id/batchText"
85          android:layout_width="0dp"
86          android:layout_height="wrap_content"
87          android:layout_weight="1"
88          android:gravity="center"
89          android:textColor="#000000" />
90  <!-- 考试类别 -->
91  <TextView
92          android:id="@+id/categoryText"
93          android:layout_width="0dp"
94          android:layout_height="wrap_content"
95          android:layout_weight="1"
96          android:gravity="center"
97          android:textColor="#000000" />
98  </LinearLayout>
99  </TableLayout>
```

4.4.2 用 SharedPreferences 实现个人信息存储

SharedPreferences 可以实现将轻量级数据以 XML 文件存储在 Android 手机本地，方便下次程序打开上次记录数据。"豹考通"把用户填写的高考成绩、高考排名、生源地、意向省份、录取批次和考生类别通过 SharedPreferences 存储在 Android 本地，如图 4-11 所示。在实现个人信息存储前，要先在主类 SelfInfoActivity 中定义常量和获取控件 id。

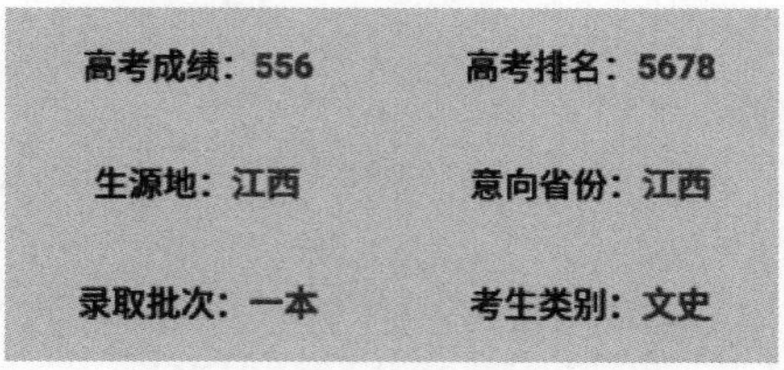

图 4-11 SharedPreferences 存储个人信息

定义常量的代码如程序清单 4-30 所示。

程序清单 4-30：Code040401\app\src\main\java\cn\jxufe\iet\code040401\SelfInfoActivity.java

```
1   private Spinner categorySpinner,batchSpinner,sourceAreaSpinner,
    targetAreaSpinner;
2   private String[] years= new String[] { "2014", "2013", "2012", "2011","2010" };
                                                        //年份的集合
3   private String[] provinces =new String[]{"上海","浙江","江西","北京"};
                                                        //省份的集合
4   private String [] batches=new String[]{"不限","一本","二本","三本","专科",
    "提前"};                                            //批次集合
```

```
5    private String [] categories=new String[]{"不限","文史","理工","综合类","艺
     术类","体育"};                              //科类集合
6    public static int sourceAreaId =2;          //生源地 id,默认为江西
7    public static int targetAreaId=2;           //意向省份 id
8    public static int batchId = 1;              //批次 id
9    public static int categoryId = 1;           //科类 id
10   private SharedPreferences selfInfoPreferences;
11   private TextView scoreText,rankText,categoryText,batchText,sourceAreaText,
     targetAreaText;
12   private int score,rank,category,batch;      //用于记录对话框选择值
13   private Button edit;                        //便捷按钮
```

获取组件 id 的代码如程序清单 4-31 所示。

程序清单 4-31：Code040401\app\src\main\java\cn\jxufe\iet\code040401\SelfInfoActivity.java

```
1    //获取组件 id
2    scoreText=(TextView)findViewById(R.id.scoreText);
3    rankText=(TextView)findViewById(R.id.rankText);
4    categoryText=(TextView)findViewById(R.id.categoryText);
5    batchText=(TextView)findViewById(R.id.batchText);
6    sourceAreaText=(TextView)findViewById(R.id.sourceAreaText);
7    targetAreaText=(TextView)findViewById(R.id.targetAreaText);
```

实现 SharedPreferences 信息存储的代码如程序清单 4-32 所示。

程序清单 4-32：Code040401\app\src\main\java\cn\jxufe\iet\code040401\SelfInfoActivity.java

```
1    selfInfoPreferences=getSharedPreferences("selfInfo", Context.MODE_
     PRIVATE);                                    //读取配置参数
2    score=selfInfoPreferences.getInt("score", -1);      //设置分数默认值为-1
3    rank=selfInfoPreferences.getInt("rank", -1);        //设置排名默认值为-1
4    category=selfInfoPreferences.getInt("categoryId", 0);  //设置科类默认值为-1
5    batch=selfInfoPreferences.getInt("batchId", 0);     //设置批次默认值为-1
6    sourceAreaId=selfInfoPreferences.getInt("sourceAreaId", 0);
                                                     //设置生源地省份默认值为-1
7    targetAreaId=selfInfoPreferences.getInt("targetAreaId",0);
                                                     //设置目标省份默认值为-1
8    boolean isFirst=selfInfoPreferences.getBoolean("isFirst", true);
```

现在实现了将信息以 XML 的形式存储在手机本地，数据也可以从本地进行读取，但还没有把数据显示在界面上供用户观看。下面新建 show 方法，把数据的信息放在 show 方法中，具体实现代码如程序清单 4-33 所示。

程序清单 4-33：Code040401\app\src\main\java\cn\jxufe\iet\code040401\SelfInfoActivity.java

```
1    private void show(){
```

```
2       if(score!=-1){
3           scoreText.setText(Html.fromHtml("高考成绩:<b><font color='red'>"
    +score+"</font></b>"));
4       }else{
5           scoreText.setText(Html.fromHtml("高考成绩:<b><font color='red'>
    暂未提供</font></b>"));
6       }
7       if(rank!=-1){
8           rankText.setText(Html.fromHtml("高考排名:<b><font color='red'>"
    +rank+"</font></b>"));
9       }else{
10          rankText.setText(Html.fromHtml("高考排名:<b><font color='red'>暂
    未提供</font></b>"));
11      }
12      categoryText.setText(Html.fromHtml(
13  "考生类别:<b><font color='red'>"+categories[category]+"</font></b>"));
14      batchText.setText(Html.fromHtml(
15  "录取批次:<b><font color='red'>"+batches[batch]+"</font></b>"));
16      sourceAreaText.setText(Html.fromHtml(
17  "生源地:<b><font color='red'>"+provinces[sourceAreaId]+"</font></b>"));
18      targetAreaText.setText(Html.fromHtml(
19  "意向省份:<b><font color='red'>"+provinces[targetAreaId]+"</font></b>"));
20  }
```

4.4.3 个人基本信息填写对话框的实现

点击"编辑"按钮将会弹出个人基本信息填写对话框让用户输入个人信息,该对话框采用自定义布局实现,如图4-12所示。实现此对话框分为以下几个步骤。

图4-12 个人基本信息填写对话框

(1) 实现对话框布局。
(2) "编辑"按钮采用匿名内部类事件监听,调用 showDialog 方法(先空留注释,后面实现)。
(3) 编写对话框中的 Spinner 内部监听类。
(4) 实现 showDialog 方法。

自定义对话框整体采用表格布局,内部结合文本显示框、编辑框、下拉列表、垂直滚动条等控件实现。其中编辑框背景图片 border.xml 在 4.2.1 节中已经实现,Spinner 自定义布局 spinner_item_batch.xml、spinner_item_category.xml 和 spinner_item_year.xml 在 4.2.1 中也已经实现,这里不具体进行讲解。对话框自定义布局 dialog_self_info_edit.xml 的代码如程序清单 4-34 所示。

程序清单 4-34:Code040401\app\src\main\res\layout\dialog_self_info_edit.xml

```
1   <LinearLayout xmlns:android="http://schemas.android.com/apk/res/android"
2       android:layout_width="match_parent"
3       android:layout_height="match_parent"
4       android:background="#66ddccbb"
5       android:orientation="vertical">
6   <TextView
7       android:layout_width="match_parent"
8       android:layout_height="wrap_content"
9       android:background="#ccbbaa"
10      android:gravity="center"
11      android:padding="5dp"
12      android:textColor="#000000"
13      android:text="@string/selfInfoTitle"
14      android:textSize="20sp" />
15  <ScrollView
16      android:layout_width="match_parent"
17      android:layout_height="wrap_content">
18  <TableLayout
19      android:layout_width="wrap_content"
20      android:layout_height="wrap_content"
21      android:layout_gravity="center"
22      android:layout_margin="10dp"
23      android:padding="10dp">
24  <LinearLayout
25      android:layout_width="match_parent"
26      android:layout_height="wrap_content"
27      android:layout_marginBottom="15dp"
28      android:orientation="horizontal">
29  <!-- 高考分数 -->
30  <TextView
```

```
31                    android:layout_width="70dp"
32                    android:layout_height="wrap_content"
33                    android:layout_marginRight="10dp"
34                    android:text="@string/score" />
35      <EditText
36                    android:id="@+id/score"
37                    android:layout_width="wrap_content"
38                    android:layout_height="wrap_content"
39                    android:background="@drawable/border"
40                    android:hint="@string/inputScore"
41                    android:inputType="number"
42                    android:padding="5dp"
43                    android:textSize="14sp" />
44      </LinearLayout>
45      <LinearLayout
46                    android:layout_width="match_parent"
47                    android:layout_height="wrap_content"
48                    android:layout_marginBottom="15dp"
49                    android:orientation="horizontal">
50      <!-- 高考排名 -->
51      <TextView
52                    android:layout_width="70dp"
53                    android:layout_height="wrap_content"
54                    android:layout_marginRight="10dp"
55                    android:text="@string/rank" />
56      <EditText
57                    android:id="@+id/rank"
58                    android:layout_width="wrap_content"
59                    android:layout_height="wrap_content"
60                    android:background="@drawable/border"
61                    android:hint="@string/rank"
62                    android:inputType="number"
63                    android:padding="5dp"
64                    android:textSize="14sp" />
65      </LinearLayout>
66      <TableRow android:layout_marginBottom="10dp">
67      <!-- 生源地 -->
68      <TextView
69                    android:layout_width="wrap_content"
70                    android:layout_height="wrap_content"
71                    android:text="@string/sourceArea" />
72      <Spinner
73                    android:id="@+id/sourceAreaSpinner"
74                    android:layout_width="wrap_content"
```

```
75                          android:layout_height="wrap_content"
76                          android:background="@drawable/border" />
77      </TableRow>
78      <TableRow android:layout_marginBottom="10dp">
79      <!-- 批次 -->
80      <TextView
81                          android:layout_width="wrap_content"
82                          android:layout_height="wrap_content"
83                          android:text="@string/batch" />
84      <Spinner
85                          android:id="@+id/batchSpinner"
86                          android:layout_width="wrap_content"
87                          android:layout_height="wrap_content"
88                          android:background="@drawable/border" />
89      </TableRow>
90      <TableRow android:layout_marginBottom="10dp">
91      <!-- 类别 -->
92      <TextView
93                          android:layout_width="wrap_content"
94                          android:layout_height="wrap_content"
95                          android:text="@string/category" />
96      <Spinner
97                          android:id="@+id/categorySpinner"
98                          android:layout_width="wrap_content"
99                          android:layout_height="wrap_content"
100                         android:background="@drawable/border" />
101     </TableRow>
102     <TableRow android:layout_marginBottom="10dp">
103     <!-- 意向省市 -->
104     <TextView
105                         android:layout_width="70dp"
106                         android:layout_height="wrap_content"
107                         android:layout_marginRight="10dp"
108                         android:text="@string/targetArea" />
109     <Spinner
110                         android:id="@+id/targetAreaSpinner"
111                         android:layout_width="wrap_content"
112                         android:layout_height="wrap_content"
113                         android:background="@drawable/border" />
114     </TableRow>
115     </TableLayout>
116     </ScrollView>
117     </LinearLayout>
```

"编辑"按钮的具体实现代码如程序清单 4-35 所示。

程序清单 4-35（部分）：Code040401\app\src\main\java\cn\jxufe\iet\code040401\SelfInfoActivity.java

```
1    edit=(Button)findViewById(R.id.edit);
2    edit.setOnClickListener(new View.OnClickListener() {
3        @Override
4        public void onClick(View v) {
5            showDialog();                        //调用对话框方法
6        }
7    });
```

在上面调用的 showDialog() 的方法还未实现，在显示的对话框方法中需要添加对话框标题、自定义 View、确定按钮。其中，需要将用户在对话框中选择列表项，所获取的信息通过 SharedPreferences 存入手机的本地 XML 文件中，因此需要实现事件监听器对 4 个 Spinner 列表进行监听。事件监听内部类的具体实现代码如程序清单 4-36 所示。

程序清单 4-36（部分）：Code040401\app\src\main\java\cn\jxufe\iet\code040401\SelfInfoActivity.java

```
1    private class MyItemSelectedListener implements AdapterView
     .OnItemSelectedListener {
2        @Override
3        public void onItemSelected(AdapterView<?> parent, View view,
4                                   int position, long id) {
5            switch (parent.getId()) {
6                case R.id.sourceAreaSpinner:
                                       //如果是选择生源地，则保存省份的序号，从 0 开始
7                    sourceAreaId = position;
8                    break;
9                case R.id.categorySpinner:       //如果是选择文理科
10                   category=position;
11                   break;
12               case R.id.batchSpinner:          //如果是选择批次
13                   batch = position;
14                   break;
15               case R.id.targetAreaSpinner:     //如果是选择意向省份
16                   targetAreaId=position;
17                   break;
18               default:
19                   break;
20           }
21       }
```

下面实现 showDialog() 方法，在 showDialog() 方法中添加对话框标题、自定义 View 和确定按钮。自定义 View 中包括将自定义布局转换成 View 对象，获取组件并定义列表适配器实现列表，同时对 4 个列表注册事件监听器。在对话框确定按钮方法中获取用户

输入和选择的信息,通过 SharedPreferences 存入手机本地 XML 文件中,具体代码如程序清单 4-37 所示。

程序清单 4-37(部分):Code040401\app\src\main\java\cn\jxufe\iet\code040401\SelfInfoActivity.java

```
1   public void showDialog(){                                    //弹出对话框
2       AlertDialog.Builder builder=new AlertDialog.Builder(this);
3       builder.setTitle("个人基本信息填写");
4       View view=this.getLayoutInflater().inflate(R.layout.dialog_self_info
          _edit, new LinearLayout(this));
5       sourceAreaSpinner = (Spinner) view.findViewById(R.id.sourceAreaSpinner);
6       categorySpinner = (Spinner) view.findViewById(R.id.categorySpinner);
7       batchSpinner = (Spinner)view.findViewById(R.id.batchSpinner);
8       targetAreaSpinner = (Spinner) view.findViewById(R.id.targetAreaSpinner);
9       final EditText scoreText = (EditText) view.findViewById(R.id.score);
                                                                 //高考成绩
10      final EditText rankText=(EditText)view.findViewById(R.id.rank);
                                                                 //高考排名
11      if(score!=-1){
12          scoreText.setText("");
13      }
14      if(rank!=-1){
15          rankText.setText("");
16      }
17      ArrayAdapter<String> areaAdapter = new ArrayAdapter<String>(
18              this, R.layout.spinner_item_year, provinces);
19      sourceAreaSpinner.setAdapter(areaAdapter);
20      targetAreaSpinner.setAdapter(areaAdapter);
21      sourceAreaSpinner.setSelection(sourceAreaId);
22      targetAreaSpinner.setSelection(targetAreaId);
23      ArrayAdapter<String> categoryAdapter = new ArrayAdapter<String>(
24              this, R.layout.spinner_item_category, categories);
25      categorySpinner.setAdapter(categoryAdapter);
26      categorySpinner.setSelection(category);                  //类型默认为理工
27      ArrayAdapter<String> batchAdapter = new ArrayAdapter<String>(
28              this, R.layout.spinner_item_batch, batches);
29      batchSpinner.setAdapter(batchAdapter);
30      batchSpinner.setSelection(batch);                        //批次默认为二本
31  MyItemSelectedListener myItemSelectedListener = new MyItemSelectedListener();
                                                                 //实例化事件监听器
32      //对 Spinner 列表进行注册
33      sourceAreaSpinner.setOnItemSelectedListener(myItemSelectedListener);
34      categorySpinner.setOnItemSelectedListener(myItemSelectedListener);
35      batchSpinner.setOnItemSelectedListener(myItemSelectedListener);
```

```
36        targetAreaSpinner.setOnItemSelectedListener(myItemSelectedListener);
37        builder.setView(view);
38        builder.setPositiveButton("保存", new DialogInterface
          .OnClickListener() {
39            @Override
40            public void onClick(DialogInterface dialog, int which) {
41                SharedPreferences.Editor editor=selfInfoPreferences.edit();
42                editor.putBoolean("isFirst", false);
43                String scoreget=scoreText.getText().toString().trim();
                                                                            //获取输入分数
44                if("".equals(scoreget)){
45                    editor.putInt("score", 0);
46                }else{
47                    editor.putInt("score", Integer.parseInt(scoreget));
                                                                            //保持编辑分数
48                    score=Integer.parseInt(scoreget);
49                }
50                String rankget=rankText.getText().toString().trim();
51                if("".equals(rankget)){
52                    editor.putInt("rank", 0);            //如果排名没填,设置排名值为 0
53                }else{
54                    editor.putInt("rank", Integer.parseInt(rankget));
                                                                            //保存分数排名
55                    rank=Integer.parseInt(rankget);
56                }
57                editor.putInt("categoryId", categoryId);     //类别 id
58                editor.putInt("batchId",batchId);            //批次 id
59                editor.putInt("sourceAreaId", sourceAreaId); //生源地 id
60                editor.putInt("targetAreaId",targetAreaId);  //意向省市 id
61                editor.commit();
62                show();
63            }
64        });
65        builder.create().show();                             //创建并显示对话框
66    }
```

如果想查看 SharedPreferences 中的内容,可以使用 Android Studio 自带的工具 Device File Explorer,可以在 Android Studio 的右侧侧边栏中找到它,如图 4-13 所示。

打开 Device File Explorer,通过目录 data/data/cn/jxufe/iet/code040401/shared_prefs 找到存储文件,如图 4-14 所示。

选中存储文件后,选择 Save As 选项,即可导出指定文件,如图 4-15 所示。

导出后,可以用记事本打开 XML 文件,查看其中的内容,如图 4-16 所示。

第 4 章　Android 客户端设计和实现

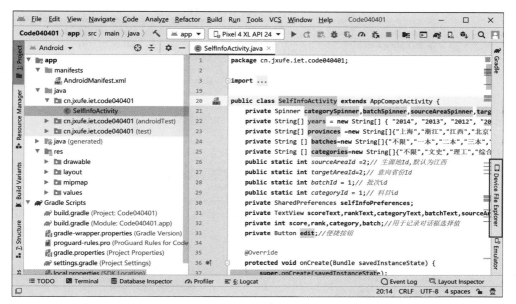

图 4-13　使用 Android Device Monitor 工具

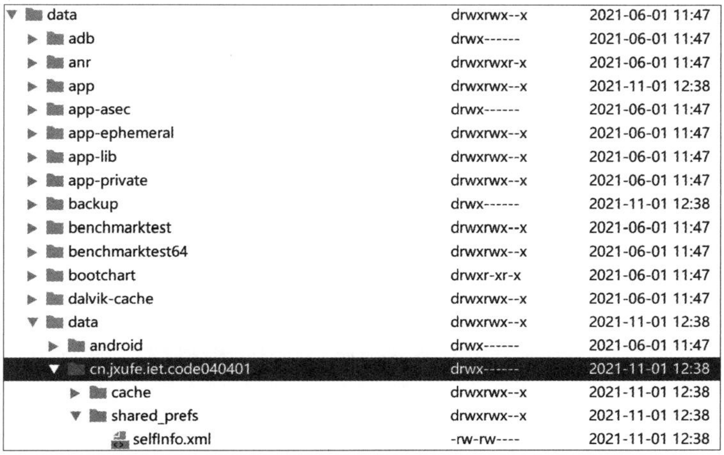

图 4-14　找到存储文件

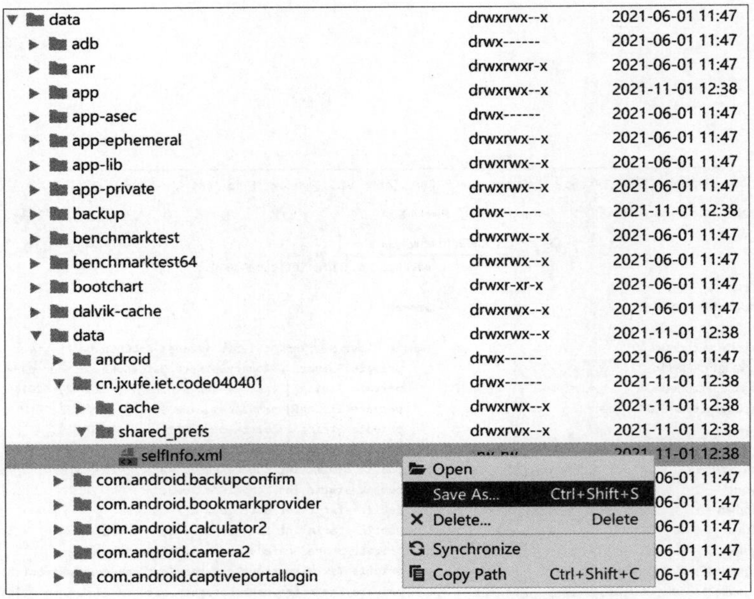

图 4-15　导出文件

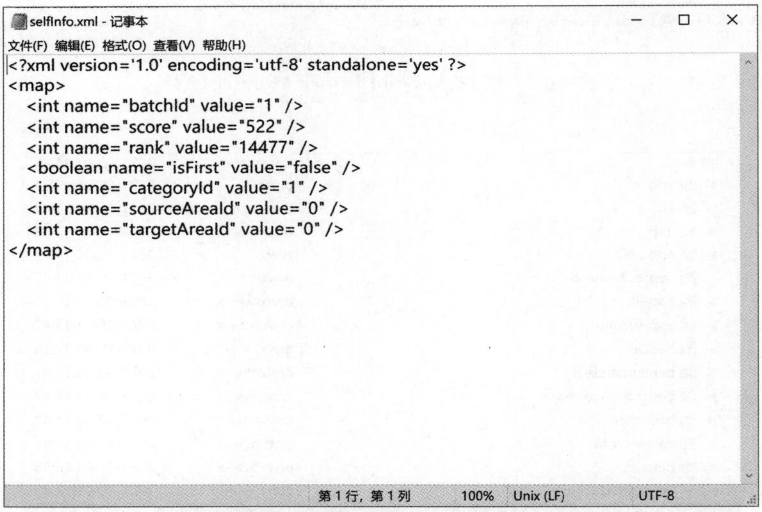

图 4-16　查看 XML 文件内容

4.5　多页面切换效果设计与实现

本节学习如何在图 4-17 所示的界面中为下方的每个选项配置一个独立的页面,实现在多个页面间的切换功能。首先是为"第一页"填充一个简单页面(见图 4-18),体会添加一个页面的基本方法。然后以"豹考通"的实际项目为例,完成多个页面的填充,以便于实现切换功能(见图 4-19)。

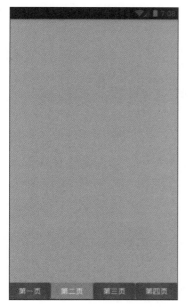

图 4-17　空选项卡的配置

图 4-18　选项卡填充页面

图 4-19　实际项目填充选项卡

4.5.1　多页面切换框架的实现

为了实现多页面切换,可将整个过程分为三步:TabHost界面布局文件设计,为选项卡添加样式,初始化选项卡设计。

1. TabHost 界面布局文件设计

在"豹考通"教学版的选项卡界面中,我们需要使用一种全新的 TabHost 布局。也就是说在整个布局的外围,我们需要使用一个＜TabHost＞＜/TabHost＞标签包裹住整个页面的内容,代码如程序清单 4-38 所示。

程序清单 4-38:Code040501\app\src\main\res\layout\activity_main.xml

```
1   <TabHost xmlns:android="http://schemas.android.com/apk/res/android"
2       android:id="@+id/mTabHost"              //为控件命名
3       android:layout_width="match_parent"     //设置宽度为 100%
4       android:layout_height="match_parent"    //设置高度为 100%
5       android:background="#ccbbaa">           //设置背景颜色
6   </TabHost>
```

整体的布局完成后,我们就可以在这个布局中添加那些页面中要用到的内容。首先,我们可以分析页面中有哪些要素(见图 4-20)。

图 4-20　主界面整体框架

直观地看,我们可以发现在这个页面中存在两大块内容,分别是由红色方框括起来的内容区和蓝色方框括起来的选项卡区域。这两个元素之间存在一个垂直分布的关系,因此我们首先需要添加一个垂直的线性布局(见程序清单 4-39)。

程序清单 4-39:Code040501\app\src\main\res\layout\activity_main.xml

```
1   <TabHost xmlns:android="http://schemas.android.com/apk/res/android"
```

```
2      android:id="@+id/mTabHost"                      //为控件命名
3      android:layout_width="match_parent"             //设置宽度为100%
4      android:layout_height="match_parent"            //设置高度为100%
5      android:background="#ccbbaa">                   //设置背景颜色
6      <LinearLayout
7         android:layout_width="match_parent"
8         android:layout_height="match_parent"
9         android:orientation="vertical">              //布局中的内容设置为垂直排布
10     </LinearLayout>
11  </TabHost>
```

接下来在线性布局中填充页面内容和选项卡。在"豹考通"教学版中，页面内容模块使用了FrameLayout(帧布局)，而选项卡控件的标签为<TabWidget></TabWidget>，代码如程序清单4-40所示。

程序清单 4-40：Code040501\app\src\main\res\layout\activity_main.xml

```
1   <?xml version="1.0" encoding="utf-8"?>
2   <TabHost xmlns:android="http://schemas.android.com/apk/res/android"
3      android:id="@+id/mTabHost"
4      android:layout_width="match_parent"
5      android:layout_height="match_parent"
6      android:background="#ccbbaa">
7   <LinearLayout
8      android:layout_width="match_parent"
9      android:layout_height="match_parent"
10     android:orientation="vertical">
            <!-- 显示单个页面的信息,这里使用了帧布局 --!>
11  <FrameLayout
12       android:id="@android:id/tabcontent"
13       //添加系统 id:tabcontent
14       android:layout_width="match_parent"
15       android:layout_height="0dp"
16       android:layout_weight="1">
17       //设置该模块填满页面剩下的位置
18  <FrameLayout
19        android:id="@+id/content"
20        android:layout_width="match_parent"
21        android:layout_height="match_parent" />
22  </FrameLayout>
            <!-- 底部的选项卡 --!>
23  <TabWidget
24       android:id="@android:id/tabs"
25       //添加系统 id:tabs
26       android:layout_width="match_parent"
```

```
27          android:layout_height="wrap_content"
28          android:background="#66666666"></TabWidget>
29    </LinearLayout>
30 </TabHost>
```

此时,我们发现 activity_main.xml 视图界面如图 4-21 所示。

注意:如果要使用 TabHost,必须在布局文件中给 TabWidget 定义名为 tabs 的 id。除此之外,还需要给布局中的一个 FrameLayout 定义一个名为 tabcontent 的 id。这两个 id 是 Android 系统自带的,如果不为这两个元素添加系统自带的 id,选项卡就无法正常使用。

2. 为选项卡添加样式

选项卡的基本布局文件创建完毕后,我们还需要再建立一个 XML 文件,用来作为选项卡的样式。在这个样式文件中,我们可以定义选项卡的宽、高、背景色、文字颜色、点击效果等内容。如果不使用样式,选项卡的样子会变得非常丑陋,甚至影响用户的正常使用。为此,我们先建立一个 tab.xml 文件用于设计选项卡的样式。找到 layout 文件夹,通过右键建立一个 Layout resource file 文件并命名为 tab,如图 4-22 所示。

图 4-21　activity_main 视图界面

接着,我们在 tab.xml 文件中写入如程序清单 4-41 所示的代码。

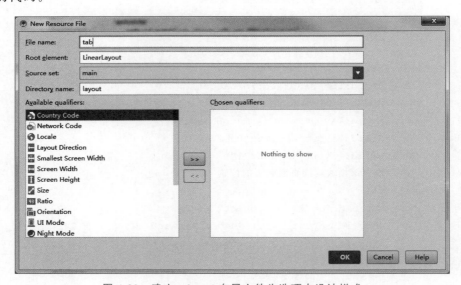

图 4-22　建立 tab.xml 布局文件为选项卡设计样式

程序清单 4-41：Code040501\app\src\main\res\layout\tab.xml

```
1   <TextView xmlns:android="http://schemas.android.com/apk/res/android"
2       android:id="@+id/title"
3       android:layout_width="0dp"
4       android:layout_weight="1"                //设置每个选项卡平分宽度
5       android:layout_height="wrap_content"
6       android:background="@drawable/tab"       //使用 drawable 样式
7       android:gravity="center"                 //设置文字居中显示
8       android:paddingBottom="8dp"              //下方内边距设置为 8
9       android:paddingTop="8dp"                 //上方内边距设置为 8
10      android:textColor="#ffffff"              //设置文字为白色
11      android:textSize="16sp"/>                //设置文字大小
```

在输入完代码后，会发现第 6 行 android：background＝"@drawable/tab"报错，这是因为我们还未在 drawable 文件夹中添加相关配置文件造成的。drawable 文件夹通常用来存放一些写好的控件特效。在"豹考通"教学版的选项卡模块中，我们在 drawable 文件夹中使用了 3 个 XML 文件来编写选项卡的特效，它们分别是 tab.xml（与 layout 文件夹下的布局文件重名，事实上这个名称可以随意设置）、tab_selected.xml 和 tab_unselected.xml。

其中，tab_selected.xml 表示的是当选项卡被选中时呈现出的样式，tab_unselected.xml 表示的是当选项卡未被选中时呈现的样式。tab.xml 文件整合了前两个文件，用于设置当选项卡处于不同状态下应该加载哪种布局文件。

我们只需要在 drawable 文件夹中分别建立程序清单 4-42～程序清单 4-44 所示的三个布局文件即可。

程序清单 4-42：Code040501\app\src\main\res\drawable\tab.xml

```
1   <?xml version="1.0" encoding="utf-8"?>
    <!--selector 标签经常用来操作控件的背景--!>
2   <selector xmlns:android="http://schemas.android.com/apk/res/android">
    <!--当控件被按下时,将控件样式设置为 tab_selected.xml--!>
3   <item android:state_pressed="true"    android:drawable="@drawable/tab_selected"></item>
    <!--当控件被选中时,将控件样式设置为 tab_selected.xml--!>
4   <item android:state_selected="true" android:drawable="@drawable/tab_selected"></item>
    <!--当控件未被按下时,将控件样式设置为 tab_unselected.xml--!>
5   <item android:state_pressed="false" android:drawable="@drawable/tab_unselected"></item>
    <!--当控件未被选中时,将控件样式设置为 tab_unselected.xml--!>
6   <item android:state_selected="false" android:drawable="@drawable/tab_unselected"></item>
7   </selector>
```

程序清单 4-43：Code040501\app\src\main\res\drawable\tab_selected.xml

```
1   <?xml version="1.0" encoding="utf-8"?>
2   <shape xmlns:android="http://schemas.android.com/apk/res/android"
3   android:shape="rectangle"><!--设置控件为矩形--!>
4   <corners android:radius="5dp"/><!--设置控件的圆角弧度--!>
5   <solid android:color="#aaaaaa"/><!-- 填充的颜色 -->
6   </shape>
```

程序清单 4-44：Code040501\app\src\main\res\drawable\tab_unselected.xml

```
1   <?xml version="1.0" encoding="utf-8"?>
2   <shape xmlns:android="http://schemas.android.com/apk/res/android"
3   android:shape="rectangle">
4   <corners android:radius="5dp"/>
5   <solid android:color="#000000"/><!-- 填充的颜色 -->
6   <gradient <!--用于设置颜色的过渡--!>
7       android:startColor="#aa444444"
8       android:centerColor="#aa666666"
9       android:endColor="#aa444444"
10      android:type="linear"/>
11  </shape>
```

当我们将这几个 XML 文件编写完后，可以发现 layout/tab.xml 布局文件中已经不再报错。

3. 初始化选项卡设计

我们已经在布局文件中完成了各个控件的插入，但此时还不能看到选项卡的效果，因为我们并没有为选项卡加载内容。接下来我们进入 MainActivity.java 文件，正式为选项卡设置内容。

首先，我们需要为每个选项卡设置标签（tag）和标题（title）。tag 指代的是选项卡每个子项的名称，title 指代的是选项卡每个子项的内容。"豹考通"教学版中的选项卡有 4 个选项，我们分别为 tag 和 title 建立一个数组来存储信息，同时要初始化选项卡控件。代码如程序清单 4-45 所示。

程序清单 4-45：Code040501\app\src\main\java\cn\jxufe\iet\MainActivity.java

```
1
2   public class MainActivity extends Activity {
    <!--初始化选项卡控件--!>
3       private TabHost mTabHost;
    <!--选项卡的标签--!>
4       private String[] tags=new String[]{"page1","page2","page3","page4"};
    <!--选项卡的内容--!>
```

```
5        private String[] titles=new String[]{"第一页","第二页","第三页","第四页"};
6    }
```

上述基本准备工作完成后,我们需要了解一个新的概念——Fragment。Fragment 是 Android 3.0 引入的新 API,译为"片段",我们可以将它理解为 Activity 中的片段或子模块。Fragment 必须被嵌入 Activity 中使用,在同一个 Activity 中可以包含多个 Fragment。

在"豹考通"教学版中,我们使用了"TabHost+Fragment"技术,用户通过单击选项卡完成页面的切换,在切换过程中,整个 Activity 并没有发生跳转,实际上发生变化的只是 Activity 中的 Fragment 部分(如图 4-23 所示)。

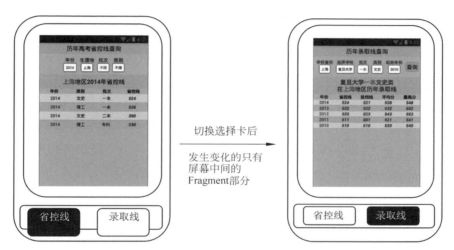

图 4-23　选项卡效果展示

如果希望使用 Fragment 来完成不同选项卡之间内容的跳转,我们需要首先让 MainActivity 集成 FragmentActivity,并且声明 onCreate()方法,代码如程序清单 4-46 所示。

程序清单 4-46:Code040502\app\src\main\java\cn\jxufe\iet\MainActivity.java

```
1         public class MainActivity extends FragmentActivity {
   <!--初始化选项卡控件--!>
2            private TabHost mTabHost;
3            <!--选项卡的标记--!>
4            private String[] tags=new String[]{"page1","page2","page3","page4"};
   <!--选项卡的内容--!>
5            private String[] titles=new String[]{"第一页","第二页","第三页","第四页"};
6            <!--onCreate()方法--!>
7            protected void onCreate(Bundle savedInstanceState) {
8                super.onCreate(savedInstanceState);
9            }
10       }
```

（此处要继承 FragmentActivity）

接下来,我们为每个选项卡填充内容,代码如程序清单 4-47 所示。

程序清单 4-47:Code040502\app\src\main\java\cn\jxufe\iet\MainActivity.java

```
1    public class MainActivity extends FragmentActivity {
2    private TabHost mTabHost;
3    private String[] tags=new String[]{"page1","page2","page3","page4"};
4    private String[] titles=new String[]{"第一页","第二页","第三页","第四页"};
5    protected void onCreate(Bundle savedInstanceState) {
6      super.onCreate(savedInstanceState);
7      requestWindowFeature(Window.FEATURE_NO_TITLE);         //设置标题栏不显示
8      setContentView(R.layout.activity_main);
9      mTabHost=(TabHost) findViewById(R.id.mTabHost);        //获取选项卡控件
10     mTabHost.setup();                                      //加载选项卡控件
11     for(int i=0;i<titles.length;i++){                      //循环添加各个选项卡
              <!--创建一个选项并为其设置标记--!>
12       TabHost.TabSpec tabSpec=mTabHost.newTabSpec(tags[i]);
              <!--加载选项卡的样式 tab.xml--!>
13       View view=getLayoutInflater().inflate(R.layout.tab, null);
14       TextView tv1=(TextView) view.findViewById(R.id.title);
         <!--设置选项卡的标题(例如第一页、第二页、省控线、录取线等)--!>
15       tv1.setText(titles[i]);
16       tabSpec.setIndicator(view);
              <!--将 id 为 content 的控件设置为选项卡对应的内容--!>
17       tabSpec.setContent(R.id.content);
              <!--将选项添加到选项卡中--!>
18       mTabHost.addTab(tabSpec);
19     }
              <!--设置选项卡默认选中第二个页面--!>
20     mTabHost.setCurrentTab(1);
21   }
```

此时,通过使用模拟器运行项目后,看到效果如图 4-24 所示。

4.5.2 为选项卡添加对应内容

1. 创建空 Fragment 布局文件及其加载类

在 4.5.1 节中,我们已经完成了 TabHost 选项卡的配置,这几个选项卡分别为"第一页""第二页""第三页""第四页"。按理说,这 4 个选项卡应该各对应一个页面,每当选项卡发生切换时,页面也会随着切换。

但我们并没有为每个选项卡设置对应的内容,接下来就准备完成每个页面。首先,我们需要建立 4 个布局文件,这 4 个布局文件分别对应 4 个选项卡,当某个选项卡被选中时,其对应的布局文件也会被加载进来。我们找到 layout 文件夹,在其中建立一个 XML 布局文件,并将其命名为 page1,如图 4-25 所示。

第 4 章 Android 客户端设计和实现

图 4-24 选项卡效果图

图 4-25 建立 page1.xml 布局文件

为简化操作，在 page1.xml 中，我们只需要添加一个位于页面正中心的文本即可，代码如程序清单 4-48 所示。

程序清单 4-48：Code040502\app\src\main\res\layout\page1.xml

```
1   <?xml version="1.0" encoding="utf-8"?>
2   <LinearLayout xmlns:android="http://schemas.android.com/apk/res/android"
3       android:layout_width="match_parent"
```

```
4       android:layout_height="match_parent"
5       android:gravity="center_vertical|center"//水平垂直居中
6       android:orientation="vertical">
7    <TextView
8       android:id="@+id/tv1"
9       android:layout_width="wrap_content"
10      android:layout_height="wrap_content"
11      android:text="这是第一页"/>
12  </LinearLayout>
```

此时，我们可以在视图界面看到效果如图 4-26 所示。

以此类推，我们继续建立另外 3 个视图界面，分别命名为 page2.xml、page3.xml、page4.xml。它们的代码与 page1.xml 代码类似，只是线性布局中添加了 TextView 控件。由于代码的重复性，page2.xml、page3.xml 和 page4.xml 在此不作展示。

4 个页面已经建立完毕，那么是不是代表我们已经可以直接在 MainActivity 中设置选项卡点击事件，并且让每个选项卡在被点击时加载这些布局文件呢？这当然是不可行的，因为我们在选项卡技术中同时使用了 Fragment 技术，用户在单击选项卡的过程中实际切换的是每一个不同的 Fragment，而这每个不同的 Fragment 又加载不同的 XML 布局文件。因此，我们需要先为每个 XML 布局文件建立一个 Java 类并让 Java 类去加载它们。

上述概念理解起来可能存在一定的难度，我们可以通过图 4-27 来帮助理解。

图 4-26　page1.xml 页面效果

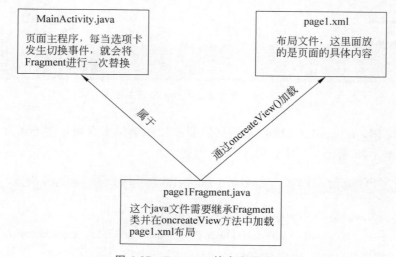

图 4-27　Fragment 技术实现流程

我们接下来分别为 page1.xml～page4.xml 编写对应的 Java 文件。为了使整个项目的目录结构更清楚，我们可以单独建一个包来存储这些 Java 文件。右击包 java.cn.jxufe.iet，新建一个 Package 文件夹并将它命名为 cn.jxufe.iet.fragment，如图 4-28 所示。

提示：建包时不要选错文件夹，我们要选择的是 main 文件夹下的 cn.jxufe.iet.fragment。

图 4-28　建立一个 fragment 包

现在我们可以看到，在 cn.jxufe.iet 文件夹下多出了一个 fragment 包。在 fragment 包中建立一个 Java 程序，命名为 page1Fragment（此处命名并无特殊要求，但为了使文件内容清楚明了，建议一般用这种格式命名）。page1Fragment 用来加载 page1.xml 布局文件，它需要首先继承 Fragment 并生成 oncreateView() 方法，代码如程序清单 4-49 所示。

程序清单 4-49：Code040502\app\src\main\java\cn\jxufe\iet\fragment\page1Fragment.java

```
1    public class page1Fragment extends Fragment {
2        public View onCreateView(LayoutInflater inflater,ViewGroup container, Bundle
3        savedInstanceState) {
    <!--加载page1.xml布局文件--!>
4            View view=inflater.inflate(R.layout.page1,container,false);
5            return view;
6        }
7    }
```

然后创建 page2Fragment.java、page3Fragment.java 和 page4Fragment.java，分别继承 Fragment，在 onCreateView 方法中加载对应布局。因为这些代码与 page1Fragment.java 比较类似，在此不作展示。

建好了 4 个 Java 文件后，文件夹项目结构如图 4-29 所示。

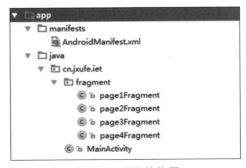

图 4-29　项目结构图

2. 通过事件监听实现多页面切换

接下来要在主程序 MainActivity 中设置一个方法来调用刚刚写好的这些 Java 文件。在"豹考通"教学版中,每当用户切换选项卡后,页面也会跟着切换到当前被选定的选项卡所对应的界面,这看上去很像一个点击事件。

事实上,这个事件是 TabHost 控件特有的,用来监听选项卡被切换时的事件。Android 中将之写作 OnTabChanged 事件,我们在 MainActivity 中使用接口来实现这个事件的监听,代码如程序清单 4-50 所示。

程序清单 4-50:Code040502\app\src\main\java\cn\jxufe\iet\MainActivity.java

```
1    public class MainActivity extends FragmentActivity {
2      private TabHost mTabHost;
3      private String[] tags=new String[]{"page1","page2","page3","page4"};
4      private String[] titles=new String[]{"第一页","第二页","第三页","第四页"};
5      protected void onCreate(Bundle savedInstanceState) {
6        super.onCreate(savedInstanceState);
7        requestWindowFeature(Window.FEATURE_NO_TITLE);
8        setContentView(R.layout.activity_main);
9        mTabHost=(TabHost) findViewById(R.id.mTabHost);
10       mTabHost.setup();
11       for(int i=0;i<titles.length;i++){
12         TabHost.TabSpec tabSpec=mTabHost.newTabSpec(tags[i]);
13         View view=getLayoutInflater().inflate(R.layout.tab, null);
14         TextView tv1=(TextView) view.findViewById(R.id.title);
15         tv1.setText(titles[i]);
16         tabSpec.setIndicator(view);
17         tabSpec.setContent(R.id.content);
18         mTabHost.addTab(tabSpec);
19       }
20       <!--为选项卡设置 OnTabChanged 监听--!>
21       mTabHost.setOnTabChangedListener(new MyTabChangedListener());
22       <!--设置选项卡默认页面为第二页,括号内为 0 时表示第一页--!>
23       mTabHost.setCurrentTab(1);
24     }
25     <!--实现 mTabHost 监听事件的具体业务逻辑--!>
26     private class MyTabChangedListener implements TabHost.OnTabChangeListener {
27       public void onTabChanged(String tabId) {
28       }
29     }
30   }
```

在 onTabChanged(String tabId)方法中有一个名为 tabId 的参数,这个参数对应的便是我们之前为每个选项卡设置的 tag。当选项卡选中"第一页"时,tag 的值会变成 page1;

而当 tag 的值变成 page1 后,我们应该在事件中让程序显示出 page1Fragment.java 类中返回的 view 对象,这个 view 对象中加载了 page1.xml 布局文件。这看上去像是一个复杂的过程,但代码其实并不复杂(见程序清单 4-51)。

程序清单 4-51:Code040502\app\src\main\java\cn\jxufe\iet\MainActivity.java

```
1   public class MainActivity extends FragmentActivity {
2     private TabHost mTabHost;
3     private String[] tags=new String[]{"page1","page2","page3","page4"};
4     private String[] titles=new String[]{"第一页","第二页","第三页","第四页"};
5     protected void onCreate(Bundle savedInstanceState) {
6       super.onCreate(savedInstanceState);
7       requestWindowFeature(Window.FEATURE_NO_TITLE);
8       setContentView(R.layout.activity_main);
9       mTabHost=(TabHost) findViewById(R.id.mTabHost);
10      mTabHost.setup();
11      for(int i=0;i<titles.length;i++){
12        TabHost.TabSpec tabSpec=mTabHost.newTabSpec(tags[i]);
13        View view=getLayoutInflater().inflate(R.layout.tab, null);
14        TextView tv1=(TextView) view.findViewById(R.id.title);
15        tv1.setText(titles[i]);
16        tabSpec.setIndicator(view);
17        tabSpec.setContent(R.id.content);
18        mTabHost.addTab(tabSpec);
19      }
20      mTabHost.setOnTabChangedListener(new MyTabChangedListener());
21      mTabHost.setCurrentTab(1);
22    }
23    private class MyTabChangedListener implements TabHost.OnTabChangeListener {
24      public void onTabChanged(String tabId) {
25        //TODO Auto-generated method stub
26        //判断单击的是哪一个选项卡
27        if(tabId.equals("page1")){
28  <!--加载 Fragment 事务,使用 page1.xml 替换当前的页面--!>
29          getSupportFragmentManager().beginTransaction()
30            .replace(R.id.content,new page1Fragment()).commit();
31        }else if(tabId.equals("page2")){
32  <!--加载 Fragment 事务,使用 page2.xml 替换当前的页面--!>
33          getSupportFragmentManager().beginTransaction()
34            .replace(R.id.content,new page2Fragment()).commit();
34        }else if(tabId.equals("page3")){
35          getSupportFragmentManager().beginTransaction()
    <!--加载 Fragment 事务,使用 page3.xml 替换当前的页面--!>
```

```
36            .replace(R.id.content,new page3Fragment()).commit();
37        }else if(tabId.equals("page4")){
<!--加载 Fragment 事务,使用 page4.xml 替换当前的页面--!>
38          getSupportFragmentManager().beginTransaction()
39            .replace(R.id.content,new page4Fragment()).commit();
40        }
41      }
42    }
43  }
```

我们注意到,在 replace() 方法中出现了两个熟悉的参数,分别是 id 为 content 的控件和一个我们之前建立好的 Java 类(page1Fragment～page4Fragment)。事实上,以 replace(R.id.content,new page1Fragment()).commit() 为例,你可以将它理解为新建了一个 page1Fragment 对象,这个对象中的 onCreateView 方法会返回一个 view 对象,而这个 view 对象加载的就是我们已经写好的 page1.xml 布局文件。在得到这个 view 对象后,我们直接将 view 对象填充在 R.id.Content 控件上,而 R.id.Content 控件在 activity_main.xml 布局文件中正好是用来存放内容的一个布局控件。当该方法被执行后,选项卡会进行切换。

此时,打开模拟器,运行项目可以看到效果如图 4-30 和图 4-31 所示。

图 4-30　项目启动时默认在第二页

图 4-31　选项卡切换效果

4.5.3　多页面内容填充

在之前的章节中,已经实现了多页面切换效果的框架,但我们只是以 4 个简单的模拟页面作为每个选项卡对应的内容。在这些页面中也只有一行诸如"这是第一页"的内容,接下来我们需要用之前章节已经完成的"省控线查询页面""录取线查询页面""报考咨询页面""个人中心页面"替换掉这些模拟页面。因为 4 个模块替换比较类似,所以在此以替换"省控线查询页面"为例进行展示,主要分为替换 XML 布局文件和替换 Java 类两个步骤。

1. 替换 XML 布局文件

首先,我们要用 Android Studio 导入之前已经做好的"省控线查询页面",选择 File→New→Import Project 选项,找到省控线查询页面 demo,将之导入。

在省控线查询页面 demo 中,我们直接找到 layout 文件夹下的页面主布局 activity_main.xml 文件,使用 Ctrl+A 键全选,然后将它里面的所有代码进行复制。

之后返回我们自己创建的 Code040503 项目,找到项目中的 page1.xml 布局文件,用刚刚复制的代码覆盖掉之前的所有代码。此时 page1.xml 代码如程序清单 4-52 所示。

程序清单 4-52:Code040503\app\src\main\res\layout\page1.xml

```
1   <?xml version="1.0" encoding="utf-8"?>
2   <LinearLayout xmlns:android="http://schemas.android.com/apk/res/android"
3       xmlns:tools="http://schemas.android.com/tools"
4       android:layout_width="match_parent"
5       android:layout_height="match_parent"
6       android:background="#66ddccbb"
7       android:gravity="center_horizontal"
8       android:orientation="vertical">
9       <TextView
10          android:layout_width="match_parent"
11          android:layout_height="wrap_content"
12          android:background="#ccbbaa"
13          android:gravity="center"
14          android:padding="5dp"
15          android:text="@string/controlLineTitle"
16          android:textColor="#000000"
17          android:textSize="20sp" /><!-- 标题文本 -->
18          ……
19          ………此处省略
20      <!-- 列表项标题 -->
21      <ListView
22          android:id="@+id/controlLineList"
23          android:layout_width="match_parent"
24          android:layout_height="wrap_content">
25      </ListView>
26      <!-- 结果列表 -->
27  </LinearLayout>
```

在复制完代码后,我们会发现代码中有很多像 android:text="@string/controlLine

Title"这样的代码报错,这是由于在省控线查询页面 demo 中,在 strings.xml 文件中添加了一些代码,而 android:background="@drawable/border"表示的是引用了 drawable 文件夹中的 border.xml 文件。我们现在只是复制了省控线查询页面 demo 中的 activity_main.xml 的代码,其他的一些配置文件并没有进行处理。

首先处理 strings.xml 文件,在省控线查询页面 demo(Code040202)中找到 res\values\strings.xml 文件,使用 Ctrl+A 键全选并将这些内容原封不动地覆盖掉我们自己项目中的 strings.xml 文件。复制完毕后,要注意将＜string name="app_name"＞Code040202＜/string＞改为＜string name="app_name"＞Code040503＜/string＞(见程序清单 4-53)。

程序清单 4-53：Code040503\app\src\main\res\values\strings.xml

```
1    <resources>
2    <string name="app_name">Code040503</string>
3    <string name="batch">批次</string>
4    <string name="category">类别</string>
5    <string name="year">年份</string>
6    <string name="sourceArea">生源地</string>
7    <string name="schoolArea">学校省份</string>
8    <string name="controlLine">省控线</string>
9    <string name="controlLineTitle">历年高考省控线查询</string>
10   </resources>
```

接下来完善 drawable 文件夹下的 XML 文件,我们找到省控线查询页面 demo 的 drawable 文件夹,直接选中 drawable 文件夹下的 border.xml 并复制到 Code040503 中对应的位置。在复制的过程中会弹出一个对话框,单击 OK 按钮即可(见图 4-32)。

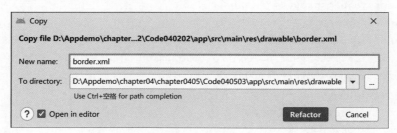

图 4-32 复制 border.xml 文件

除 strings.xml 文件和 border.xml 文件之外,我们还注意到在 Code040503 的 layout 文件夹下存在 list_item_control_line.xml、spinner_item_batch.xml、spinner_item_category.xml、spinner_item_year.xml 四个布局文件。它们的用途是为 ListView 控件和 Spinner 控件设置相关样式,在此不作详细说明。我们只需要将这 4 个文件复制到 Code040503 项目的 layout 文件夹下。

此时 Code040503 的项目结构如图 4-33 所示。

第 4 章　Android 客户端设计和实现

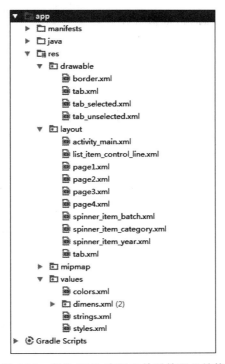

图 4-33　替换 XML 布局文件后的项目结构

2. 替换 Java 类

在上一步中我们替换了所有的 XML 配置文件，但这并不代表替换工作已经结束，Java 类的替换相较于 XML 文件更加复杂。在省控线查询页面 demo(Code040203)中，我们进入 Code040503\app\src\main\java\cn\jxuft\iet\Code040503 文件夹下，可以看到里面除了有一个页面主程序 ControlLineActivity 外，还分别有三个名为 adapter、bean、util 的包。

我们依葫芦画瓢，在 Code040501 的相应文件夹下也建立三个这样的包，如图 4-34 所示。

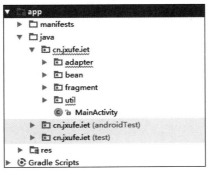

图 4-34　Code040503 包结构

103

然后我们将 Code040202 中每个包下的文件都复制到 Code040503 中对应的位置，如图 4-35 所示。

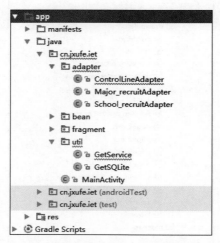

图 4-35　复制 Code040502 相应文件后的 Code040503 包结构

我们发现在复制后整个项目错误百出。不用紧张，这只是有些包未正常导入导致的，我们首先进入报错的 ControlLineAdapter 文件中修改错误。

ControlLineAdapter.java 代码如程序清单 4-54 所示。

程序清单 4-54：Code040503\app\src\main\java\cn\jxufe\iet\adapter\ControlLineAdapter.java

```
1    public class ControlLineAdapter extends BaseAdapter {
2        private Context context;
3        private List<ControlLine>controlLineList;
4    public ControlLineAdapter
5        (Context context, List<ControlLine> controlLineList) {
6        //TODO Auto-generated constructor stub
7        this.context=context;
8        this.controlLineList=controlLineList;
9        }
10   ……略
11   }
```

ControlLineAdapter.java 文件之所以会报错，很显然在于代码引用的是其他地方的地址，因此我们需要将它们删除并改成 Code040503 项目中的引用。操作过程是先将原本的引用删除，然后将鼠标指针放在报错的地方，导入类。

最后，我们要替换主程序，也就是将 Code040202 中的 ControlLineActivity 替换掉 Code040503 中的 page1Fragment，这个过程与之前不同，并不能使用简单的覆盖。我们应当注意到在 ControlLineActivity 中使用到的是 onCreate()方法，而 page1Fragment 中使用的却是 onCreateView 方法。它们不同的原因是 ControlLineActivity 类继承的是

AppCompatActivity，是一个普通的 Activity。而 page1Fragment 继承的是 Fragment，是一个位于主程序 MainActivity 中的 Fragment，它的 onCreateView 方法必须返回一个 View 类型的对象。那么，面对这种情况，我们应该如何做呢？

（1）首先将 ControlLineActivity 中定义的一些数组、控件、集合代码复制到 page1Fragment 中。不管程序如何变化，这些东西都是固定的。

（2）接下来要做的是将 ControlLineActivity 中 onCreate（）方法的内容复制到 page1Fragment 的 onCreateView（）方法中，但是我们要注意到只能从 onCreate（）方法中下面这行代码开始复制到这个方法的结束。

```
yearSpinner = (Spinner) findViewById(R.id.yearSpinner);
```

不能复制 setContentView（）方法，否则会产生冲突。

（3）代码复制完毕后，我们发现报出了非常多的错误。我们可以使用 Alt＋Enter 键解决 GetService 类和 ControlLineAdapter 类的导入错误，但是有很多其他错误却无法解决。

这些错误出现的原因是 Fragment 和 Activity 之间存在一些差异。如果一个类继承的是 Fragment，那么 Android 中会使用一个 View 类型的对象来承载这个 Fragment 中的内容，这也是 onCreateView 方法必须返回一个 View 对象的原因。我们在解决错误时，只需要将代码

```
yearSpinner = (Spinner) findViewById(R.id.yearSpinner);
```

改为

```
yearSpinner = (Spinner) view.findViewById(R.id.yearSpinner);
```

（4）在 ArrayAdapter 的构造方法中，this 应该指向的是一个 Activity 本身，page1Fragment 并不是一个 Activity，它只是依附于 Activity 中的一个 Fragment。如果此处用 this，则会指向这个 Fragment，因此代码出现了错误。

```
ArrayAdapter<String> yearAdapter = new ArrayAdapter<String>(
    this, R.layout.spinner_item_year, years);
yearSpinner.setAdapter(yearAdapter);
```

修改为

```
ArrayAdapter<String> yearAdapter = new ArrayAdapter<String>(
    this.activity(), R.layout.spinner_item_year, years);
yearSpinner.setAdapter(yearAdapter);
```

通过使用 this.getactivity（）来获得 Activity，代码就不会再报错。
ControlLineFragment.java 代码现在如程序清单 4-55 所示。

程序清单 4-55：Code040503\app\src\main\java\cn\jxufe\ief\fragment\ControlLineFragment.java

```
1   public class ControlLineFragment extends Fragment {
2       private Spinner yearSpinner, provinceSpinner, categorySpinner,
```

```java
     batchSpinner;                                          //下拉列表控件
3    private String[] years = new String[]{"2014", "2013", "2012", "2011",
     "2010"};                                               //年份的集合
4    private List<String> provinces = new ArrayList<String>();
                                                            //省份的集合
5    private String[] batchs = new String[]{"不限", "一本", "二本", "三本",
     "专科", "提前"};
6    private String[] categorys = new String[]{"不限", "文史", "理工", "综合
     类", "艺术类", "体育"};
7    private int sourceAreaId, yearId, categoryId, batchId;
8    private TextView controlLineTitle;                     //提示标题
9    private List<ControlLine> controlLineList;             //分数线的集合
10   private ListView controlLineListView;                  //显示分数线的列表
11   public View onCreateView(LayoutInflater inflater, @Nullable ViewGroup
     container, @Nullable Bundle savedInstanceState) {
12       View view = inflater.inflate(R.layout.page2, container, false);
13       yearSpinner = (Spinner) view.findViewById(R.id.yearSpinner);
                                                            //年份的下拉列表
14       provinceSpinner = (Spinner) view.findViewById(R.id.areaSpinner);
                                                            //省份的下拉列表
15       categorySpinner = (Spinner) view.findViewById(R.id.categorySpinner);
16       batchSpinner = (Spinner) view.findViewById(R.id.batchSpinner);
17       controlLineTitle = (TextView) view.findViewById(R.id.controlLineTitle);
                                                            //显示结果标题
18       controlLineListView = (ListView) view.findViewById(R.id.controlLineList);
                                                            //显示查询结果
19       ArrayAdapter<String> yearAdapter = new ArrayAdapter<String>(
20               getActivity(), R.layout.spinner_item_year, years);
21       yearSpinner.setAdapter(yearAdapter);
22       GetSQLite getSQLite = new GetSQLite();
23       provinces = getSQLite.setProvince();               //获取省份集合
24       ArrayAdapter<String> provincesAdapter = new ArrayAdapter<String>(
25               getActivity(), R.layout.spinner_item_year, provinces);
26       provinceSpinner.setAdapter(provincesAdapter);
27       ArrayAdapter<String> categoryAdapter = new ArrayAdapter<String>(
28               getActivity(), R.layout.spinner_item_category, categorys);
29       categorySpinner.setAdapter(categoryAdapter);
30       categorySpinner.setAdapter(categoryAdapter);
31       ArrayAdapter<String> batchAdapter = new ArrayAdapter<String>(
32               getActivity(), R.layout.spinner_item_batch, batchs);
33       batchSpinner.setAdapter(batchAdapter);
34       MyItemSelectedListener itemSelectedListener = new
     MyItemSelectedListener();
35       yearSpinner.setOnItemSelectedListener(itemSelectedListener);
```

```
36          provinceSpinner.setOnItemSelectedListener(itemSelectedListener);
37          categorySpinner.setOnItemSelectedListener(itemSelectedListener);
38          batchSpinner.setOnItemSelectedListener(itemSelectedListener);
39          provinceSpinner.setSelection(sourceAreaId);
40          categorySpinner.setSelection(categoryId);
41          batchSpinner.setSelection(batchId);
42      /*
43   显示省控线列表
44       */
45          GetService getService = new GetService();
46          controlLineList = getService.setControlLineList();
                                                //为controlLineList集合添加集合内容
47          ControlLineAdapter controlLineAdapter = new ControlLineAdapter
48                  (getActivity(), controlLineList);
49          controlLineListView.setAdapter(controlLineAdapter);
50          return view;
51      }
52      private class MyItemSelectedListener implements
                AdapterView.OnItemSelectedListener {
53          @Override
54          public void onItemSelected(AdapterView<?> parent, View view,
55                                     int position, long id) {
56              switch (parent.getId()) {
57                  case R.id.areaSpinner:           //如果是选择省份列表
58                      sourceAreaId = position;
59                      break;
60                  case R.id.yearSpinner:           //如果是选择年份列表
61                      yearId = position;
62                      break;
63                  case R.id.categorySpinner:       //如果是选择科类列表
64                      categoryId = position;
65                      break;
66                  case R.id.batchSpinner:          //如果是选择批次列表
67                      batchId = position;
68                      break;
69                  default:
70                      break;
71              }
72      /*
73   设置标题显示
74       */
75              controlLineTitle.setText(Html.fromHtml("<font color=red><b>"
76                      + provinces.get(sourceAreaId)
77                      + "</b></font>地区<font color=blue><b>" + years[yearId]
```

```
78                      + "</b></font>年省控线"));
79              }
80          @Override
81          public void onNothingSelected(AdapterView<?> parent) {
82          }
83      }
84  }
```

我们再运行一次项目,项目初始界面如图 4-36 所示。

当点击选项卡"第一页"后,呈现的界面如图 4-37 所示。

图 4-36　运行项目后的初始界面

图 4-37　"第一页"选项卡对应的内容

至此,省控线查询页面已经成功地被我们放在选项卡模块中,其他模块按照以上操作步骤进行操作。

4.6　绘制趋势图

在专业分数线查询页面点击按钮会生成趋势图,本节主要任务是完成趋势图的设计和实现。实现趋势图分为四步。

(1) 完成趋势图类,运行效果如图 4-38 所示。

(2) 绘制坐标轴,运行效果如图 4-39 所示。

(3) 绘制趋势线,运行效果如图 4-40 所示。

(4) 编写趋势图相关 Activity 和 Fragment,运行效果如图 4-41 所示。

图 4-38　完成趋势图类

图 4-39　绘制坐标轴

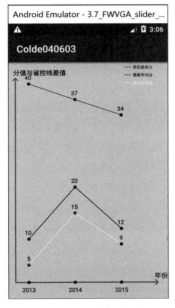
图 4-40　绘制趋势线

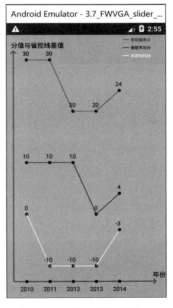
图 4-41　完成项目趋势图

4.6.1　绘制图类

在 Android 中，我们接触过很多控件，简单的如 Button、TextView，复杂的如 ListView。这些控件形态各异，能给我们提供的功能也各不相同。聪明的同学肯定会说

"'豹考通'中的趋势图肯定也是一种控件,我们只要在控件栏中找到它然后直接使用就行"。

这句话说对了一半,严格来说,趋势图确实是一种控件,但它却并不是 Android 系统自带的。趋势图控件的外观和使用方法全部都是由我们开发者自己绘制而成。一款 App 想要有各式各样极具个性的功能,只依靠系统自带的控件是很难做到的。开发者们通常都会自己着手编写一些富有特色的控件,我们称这些控件为一种"自定义的 View"。

所谓的 View 是 Android 系统中的一个超类,它是所有控件的父类。每一个 View 都有一个用于绘图的画布,我们将自定义的内容绘制在画布上,就可以变成一个自定义的 View,产生一个全新种类的控件。趋势图便是如此,我们先不急于直接编写趋势图控件,而从建立一个最简单的 View 开始。

打开 Android Studio 开发工具,选择 File→New→New Project 选项新建一个项目,将它命名为 Code040601。

现在我们建立一个属于我们自己的控件,具体的操作方法是建立一个类,然后让它继承 View 类。为了让项目结构更加清晰,我们把它单独放在一个包下。找到 cn.jxufe.iet 包,选中它并右击 New→Package 建立一个名为 View 的包,接着在这个包中建立一个普通的 Java 文件并命名为 TrendView。

文件建好以后,我们开始编写代码。前面说到,如果想要编写一个自定义的控件,首先需要让这个类继承一个 View 类,除此之外,还需要写至少两个构造方法。此时 TrendView 的代码如程序清单 4-56 所示。

程序清单 4-56:Code040601\app\src\main\java\cn\jxufe\iet\View\TrendView.java

```
1   package cn.jxufe.iet.View;
2   import android.content.Context;
3   import android.util.AttributeSet;
4   import android.view.View;
5   public class TrendView extends View {
6     public TrendView(Context context) {
7       super(context);
8     }
9     public TrendView(Context context, AttributeSet attrs) {
10      super(context, attrs);
11    }
12  }
```

找到项目中的 layout 文件夹,在里面建立一个名为 trendview 的 XML 文件。我们需要使用这个 XML 文件来引用刚刚写好的自定义控件。

程序清单 4-57 表示引用了路径 cn.jxufe.iet.View 下的 TrendView 控件,也就是我们刚刚写好的那个控件。你可能会觉得路径名的输入过于烦琐,但是在 Android Studio 中其实只需要输入 TrendView,系统就会自动提示出前面的路径名。

程序清单 4-57：Code040601\app\src\main\res\layout\trendview.xml

```
1    <cn.jxufe.iet.code040601.View.TrendView
2        android:layout_width="match_parent"
3        android:layout_height="match_parent"/>
```

此时，activity_main.xml 代码如程序清单 4-58 所示。

程序清单 4-58：Code040601\app\src\main\res\layout\activity_main.xml

```
<?xml version="1.0" encoding="utf-8"?>
<LinearLayout xmlns:android="http://schemas.android.com/apk/res/android"
    android:orientation="vertical" android:layout_width="match_parent"
    android:layout_height="match_parent">
<cn.jxufe.iet.view.TrendView
    android:layout_width="match_parent"
        android:layout_height="match_parent" />
</LinearLayout>
```

提示：因为 Android Studio 的版本不同，可能 activity_main.xml 文件中默认使用的是 RelativeLayout 布局，遇到这种情况时请将相对布局改为上述代码中的线性布局。

我们切换到 XML 的视图页面，此时显示效果如图 4-42 所示。

我们可以看到，一个硕大的空白控件已经占满整个屏幕，但这个控件并没有任何内容，因为我们还没有在 TrendView 中添加任何设置。

4.6.2　用绘图类绘制坐标轴

1. 绘制 X 轴和 Y 轴直线

使用 Paint 类设置画笔

（1）整个 TrendView 类其实就是一块画布，而我们的任务就是在这块画布上增加元素。用来绘制这些元素的东西是 Paint 类。我们需要确定画笔的颜色、粗细、样式，然后使用画笔绘制出想要的图形。

首先，我们全局定义一个画笔对象。

```
private Paint xyPaint;
```

（2）接着建立一个 init()方法并在该方法内对 xyPaint 画笔的属性进行一些设置（见程序清单 4-59）。

图 4-42　activity_main.xml 视图页面

程序清单 4-59：Code040602\app\src\main\java\cn\jxufe\iet\view\TrendView.java

```
1    public void init(){
2        xyPaint = new Paint();                          //初始化画笔
```

```
3    xyPaint.setAntiAlias(true);              //设置画笔使用抗锯齿
4    xyPaint.setStrokeWidth(3);               //设置画笔粗细
5    xyPaint.setColor(Color.RED);             //设置画笔颜色为红色
6    }
```

此时，TrendView 的代码如程序清单 4-60 所示。

程序清单 4-60：Code040602\app\src\main\java\cn\jxufe\iet\view\TrendView.java

```
1    package cn.jxufe.iet.View;
2    import android.content.Context;
3    import android.graphics.Color;
4    import android.graphics.Paint;
5    import android.util.AttributeSet;
6    import android.view.View;
7    public class TrendView extends View {
8    public TrendView(Context context) {
9    super(context);
10   }
11   public TrendView(Context context, AttributeSet attrs) {
12   super(context, attrs);
13   }
14   private Paint xyPaint;                    //定义一个画笔 Paint 对象
15   public void init(){                       //该方法用于初始化画笔设置
16   xyPaint=new Paint();
17   xyPaint.setAntiAlias(true);
18   xyPaint.setStrokeWidth(3);
19   xyPaint.setColor(Color.RED);
20   }
21   }
```

（3）现在已定义好画笔的属性，但视图界面并没有发生变化，这是因为我们还没有使用画笔在画布上进行"创作"。

创建 onDraw()方法

为了实现在画笔上进行绘制，我们需要编写一个 onDraw()方法。我们只需要在 onDraw 方法中进行一系列的操作，就可以实现控件的绘制。在 Android Studio 中，只需要输入 ondraw 这个单词，onDraw()方法就会自动提示出来。生成的 onDraw()方法如下所示：

```
protected void onDraw(Canvas canvas) {
    super.onDraw(canvas);
}
```

在 onDraw 方法中有一个 Canvas 对象，它代表的意思是"画布"。我们通过对这个对象进行操作，就可以在画布上创建许多富有个性的内容，首先绘制"豹考通"教学版的趋势图模块中的坐标轴。

坐标轴本质上就是两条直线，而 X 轴和 Y 轴的坐标轴顶端又分别有两个箭头。其实这个箭头也是由两条直线构成的，而并非直接引用一个所谓的"箭头控件"。现在我们知道了绘制的原理，但是如何才能把这些线绘制在我们想要的地方呢？

这时需要使用 canvas 对象中的 drawLine()方法，如图 4-43 所示。

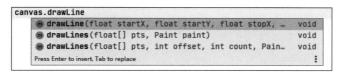

图 4-43　调用 canvas 对象中的 drawLine()方法

drawLine(float startX, float startY, float stopX, float stopY, Paint paint)方法是 canvas 对象中用来绘制一条直线的方法，它分别有如下几个参数。

- startX：起始点的 X 坐标。
- startY：起始点的 Y 坐标。
- stopX：终止端点的 X 坐标。
- stopY：终止端点的 Y 坐标。
- paint：绘制直线所使用的画笔。

在绘制直线时，我们还需要做一点准备工作，在全局中定义两个变量 width 和 height，然后在 onDraw()方法中使用 getHeight()方法和 getWidth()方法获取当前控件的高和宽并赋值给 height 和 width。通过这种方法可以较简单地得到一些坐标信息。

接下来，在 onDraw()方法中使用 drawLine()方法绘制 X 轴。代码如下所示：

```
canvas.drawLine(20, height - 50, width - 30, height - 50, xyPaint);
```

在这段代码中，初始的 X 坐标设置为 20，初始的 Y 坐标设置为控件的高度往下 50 个单位。终止端点的 X 坐标设置为控件的宽度往左 30 个单位，终止端点的 Y 坐标设置为控件的高度往下 50 个单位。绘制这条直线的画笔为 xyPaint。此时 TrendView.java 代码如程序清单 4-61 所示。

程序清单 4-61：Code040602\app\src\main\java\cn\jxufe\iet\View\TrendView.java

```
1   public class TrendView extends View {
2       public TrendView(Context context) {
3           super(context);
4       }
5       public TrendView(Context context, AttributeSet attrs) {
6           super(context, attrs);
7       }
8       private Paint xyPaint;                          //定义一个画笔 Paint 对象
9       private int width,height;
10      public void init(){                             //该方法用于初始化画笔设置
```

```
11    xyPaint=new Paint();
12    xyPaint.setAntiAlias(true);
13    xyPaint.setStrokeWidth(3);
14    xyPaint.setColor(Color.RED);
15  }
16  @Override
17  protected void onDraw(Canvas canvas) {
18    super.onDraw(canvas);
19    init();                                               //调用初始画笔的方法
20    width=this.getWidth();
21    height=this.getHeight();
22    canvas.drawLine(20, height - 50, width - 30, height - 50, xyPaint);
                                                            //绘制X轴
23  }
24  }
```

提示：不要忘记在 onDraw()方法中调用初始化画笔的方法 init()。

此时，我们进入视图页面查看 trendview.xml，可看到已经绘制了一条红色的直线（见图 4-44）。

然后绘制 Y 轴直线，同样也是使用 drawLine 方法，但方法中的参数需要改变，代码如下所示：

```
canvas.drawLine(20,height-50,20,50,xyPaint);
```

因为 Y 轴直线同样是发源于原点，所以前两个参数不变，但它是一条竖线，因此它的 X 轴终止端点和起始点是相同的 20。我们把 Y 轴直线的长度设定为 50，因此 Y 轴的终止点设置为 50。

此时 XML 视图界面如图 4-45 所示。

图 4-44　X 轴绘制完成

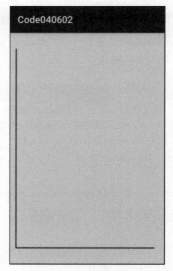
图 4-45　Y 轴绘制完成

2. 绘制坐标轴的箭头

接下来绘制坐标轴的箭头。前文说到,坐标轴的箭头实质就是两条直线,因此我们需要找到合适的位置绘制四条直线。首先绘制 X 轴的箭头。在这之前,我们需要在全局中定义一个变量 arrawSpace 来保存箭头之间的间距,如图 4-46 所示。

图 4-46　箭头的间距 arrawSpace

这里我们将 arrawSpace 设置为 8 个单位,代码如下所示:

```
private int arrawSpace=8;
```

我们使用 drawLine()方法绘制 X 轴的箭头,代码如下所示:

```
canvas.drawLine(width - 30 - arrawSpace, height - 50 - arrawSpace,
        width - 30, height - 50, xyPaint);          //绘制 X 轴箭头下半部分
canvas.drawLine(width - 30 - arrawSpace, height - 50 + arrawSpace,
        width - 30, height - 50, xyPaint);          //绘制 X 轴箭头上半部分
```

要注意的是,组成箭头的两条直线并不是发源于同一个点,而是汇聚于同一个点。返回视图页面,显示的结果如图 4-47 所示。

Y 轴箭头的绘制方法大同小异,只有坐标轴需要进行一些变动,代码如下所示:

```
canvas.drawLine(20 + arrawSpace, 50 + arrawSpace, 20, 50, xyPaint);
                                                    //绘制 Y 轴箭头右半部分
canvas.drawLine(20 - arrawSpace, 50 + arrawSpace, 20, 50, xyPaint);
                                                    //绘制 Y 轴箭头左半部分
```

此时页面效果如图 4-48 所示。

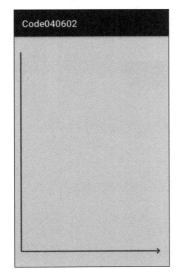

图 4-47　X 轴箭头绘制完毕

图 4-48　Y 轴箭头绘制完毕

提示:如果对 canvas 对象中坐标的空间概念比较模糊,不妨自己多使用几次

drawLine()方法并填入不同的参数进行测试。

3. 绘制坐标轴上的文字

现在绘制坐标轴上的文字。在"豹考通"教学版的趋势图中，X轴代表"年份"，Y轴代表"分值与省控线差值"。在canvas上添加文字需要使用方法drawText(String text, float x, float y, Paint paint)，该方法中的参数分别如下。

- text：要插入的文字内容。
- x：被插入文字的X坐标。
- y：被插入文字的Y坐标。
- paint：文字所使用的画笔。

为了让文字更醒目，需要创建一个新的画笔来绘制文字，因此应该在全局中定义一个文字画笔，代码如下所示：

```
private Paint textPaint;
```

然后在init()方法中初始化文字画笔textPaint，代码如下所示：

```
textPaint = new Paint();
textPaint.setAntiAlias(true);
textPaint.setColor(Color.BLACK);              //文本颜色为黑色
textPaint.setTextSize(textSize);              //文字大小为20
```

接着，我们开始在onDraw()方法中绘制文字，代码如下所示：

```
canvas.drawText("年份", width - 40, height - 50 - arrawSpace - 5,textPaint);
                                              //X轴坐标提示文本
canvas.drawText("分值与省控线差值", 5 + arrawSpace, 40, textPaint);
                                              //Y轴坐标提示文本
```

此时视图页面如图4-49所示。

图4-49 坐标轴文字绘制完毕

TrendView.java(参照 Code040602 同名文件)的代码如程序清单 4-62 所示。

程序清单 4-62：Code040602\app\src\main\java\cn\jxufe\iet\View\TrendView.java

```
1    public class TrendView extends View {
2        private Paint xyPaint;                                    //定义一个坐标轴画笔
3        private Paint textPaint;                                  //定义一个文字画笔
4        private int width,height;
5        private int arrawSpace = 8;                               //箭头的间距
6        public TrendView(Context context) {
7            super(context);
8        }
9        public TrendView(Context context, AttributeSet attrs) {
10           super(context, attrs);
11       }
12       public void init(){                                       //该方法用于初始化画笔设置
13           xyPaint=new Paint();
14           xyPaint.setAntiAlias(true);
15           xyPaint.setStrokeWidth(3);
16           xyPaint.setColor(Color.RED);
17           textPaint = new Paint();
18           textPaint.setAntiAlias(true);
19           textPaint.setColor(Color.BLACK);                      //文本颜色为白色
20           textPaint.setTextSize(20);                            //文字大小为 20
21       }
22       @Override
23       protected void onDraw(Canvas canvas) {
24           super.onDraw(canvas);
25           init();                                               //调用初始画笔的方法
26           width=this.getWidth();
27           height=this.getHeight();
28           canvas.drawLine(20, height - 50, width - 30, height - 50, xyPaint);
                                                                   //绘制 X 轴
29           canvas.drawLine(20,height-50,20,50,xyPaint);
30           canvas.drawLine(width - 30 - arrawSpace, height - 50 - arrawSpace,
31           width - 30, height - 50, xyPaint);                    //绘制 X 轴箭头
32           canvas.drawLine(width - 30 - arrawSpace, height - 50 + arrawSpace,
33           width - 30, height - 50, xyPaint);                    //绘制 X 轴箭头
34           canvas.drawLine(20 + arrawSpace, 50 + arrawSpace, 20, 50, xyPaint);
                                                                   //绘制 Y 轴箭头
35           canvas.drawLine(20 - arrawSpace, 50 + arrawSpace, 20, 50, xyPaint);
                                                                   //绘制 Y 轴箭头
36           canvas.drawText("年份", width - 40, height - 50 - arrawSpace - 5,
37           textPaint);                                           //X 轴坐标提示文本
```

```
38    canvas.drawText("分值与省控线差值", 5 + arrawSpace, 40, textPaint);
                                                            //Y轴坐标提示文本
39  }
40 }
```

4.6.3 用绘图类绘制趋势线——源数据来自数组

在上一节中,已经绘制了趋势图页面的坐标轴。坐标轴在任何情况下都是固定的,因此可以直接指定某个具体的数值作为坐标点。但坐标轴上的点和线能不能使用我们指定的坐标呢?

这当然是不合适的,因为我们的趋势图是根据"录取线查询"页面的数据绘制的,趋势图的点呈现的分布位置全部取决于从"录取线查询"页面得到的数据。南昌大学的录取线和江西财经大学的录取线信息显然是不同的,它们各自对应的趋势图也会不同,因此趋势图中的数据是动态的、变动的。

但是,从另外一个页面中获取数据需要涉及较多的内容,在本节中我们暂时使用一组模拟数据来绘制趋势图,在下一节中我们会将模拟数据换成动态数据。

1. 创建模拟数据

在"豹考通"教学版中,趋势图是根据"省控线查询"页面的数据绘制的,我们为了让大家由浅入深地理解该模块的制作,先使用一组模拟数据绘制趋势图。

模拟数据显然不能只是一个简单的变量或数组,刚刚也说到,趋势图是根据"录取线查询"页面的数据绘制的,因此我们的模拟数据也必须是一组"录取线"类型的数据。我们可以导入项目Code040503,在该项目下的bean目录下找到"录取线"的Java类,查看这个类中包含了哪些属性,如图4-50所示。

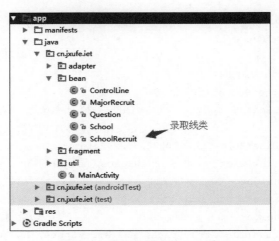

图 4-50 查询录取线的 Java 类

我们可以发现SchoolRecruit类中有许多属性,为了节约工序,直接将Code040503中的SchoolRecruit类复制到Code040603中,注意不要忘记建一个名为bean的包。此时项目结构如图4-51所示。

第 4 章　Android 客户端设计和实现

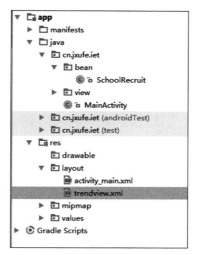

图 4-51　在 Code040603 中建立对应的包和实体类

类建好以后，我们使用一个 List 集合存储模拟的数据。首先进入 TrendView.java 文件，在全局中定义一个 SchoolRecruit 类型的集合，代码如下所示：

privateList<SchoolRecruit>schoolRecruits;

这个名为 schoolRecruits 的集合就是我们要存放模拟数据的地方。接下来写一个 setSchoolRecruits() 方法为集合赋值，为其添加一系列模拟数据，该方法的代码如程序清单 4-63 所示。

程序清单 4-63：TrendView.java 中的 setSchoolRecruits 方法

```
1  public List<SchoolRecruit>setSchoolRecruits(){
                                              //建立集合,向集合中填充模拟数据
2      schoolRecruits=new ArrayList<SchoolRecruit>();
3  SchoolRecruit sc1=new SchoolRecruit();      //建立第一个录取线对象
4  sc1.setsRecruitHScore(560);                 //设置录取最高分
5  sc1.setControlLine(520);                    //设置省控线
6  sc1.setEnterLine(525);                      //设置录取线
7      sc1.setsRecruitAScore(530);             //设置录取平均分
8  sc1.setsRecruitYear(2013);                  //设置年份
9  SchoolRecruit sc2=new SchoolRecruit();
10     sc2.setsRecruitHScore(550);
11     sc2.setControlLine(513);
12     sc2.setEnterLine(528);
13     sc2.setsRecruitAScore(533);
14     sc2.setsRecruitYear(2014);
15 SchoolRecruit sc3=new SchoolRecruit();
16     sc3.setsRecruitHScore(565);
```

```
17      sc3.setControlLine(531);
18      sc3.setEnterLine(540);
19      sc3.setsRecruitAScore(543);
20      sc3.setsRecruitYear(2015);
21      schoolRecruits.add(sc3);              //将对象加入集合中
22      schoolRecruits.add(sc2);
23      schoolRecruits.add(sc1);
24      return schoolRecruits;
25  }
```

在趋势图页面中,我们需要使用的数据有年份、录取最高分、录取平均分、省控线、录取线五个种类,在这里我们取三年的数据,因此要建立 3 个对象。这个方法有一个返回值,它会返回一个 SchoolRecruit 类型的集合,我们直接在 onDraw()方法中使用代码获取这个方法的返回值。

schoolRecruits=new ArrayList<SchoolRecruit>();   //初始化集合
schoolRecruits=setSchoolRecruits();   //向集合中填充数据

数据有了,接下来就是根据这些数据绘制趋势图中的"点"。

2. 绘制 X 轴上的信息

现在我们绘制 X 轴上的年份。接收到的数据中有多少个年份的信息,X 轴上就会显示多少个年份。我们的模拟数据中设置了 3 个年份(2013、2014、2015),与之对应的,X 轴上也应该出现 3 个点。

在此之前,我们要在全局中添加一个绘制点的画笔 pointPaint 和绘制年份文字的画笔 numPoint,代码如下所示:

private Paint pointPaint,numPanint;   //定义点画笔和年份字体画笔

然后,在 init()方法中为这两种画笔设置样式,如程序清单 4-64 所示。

程序清单 4-64:TrendView.java 中的 init()方法

```
1   public void init(){
2       xyPaint=new Paint();
3       xyPaint.setAntiAlias(true);
4       xyPaint.setStrokeWidth(3);
5       xyPaint.setColor(Color.RED);
6       //文本画笔的初始化
7       textPaint = new Paint();
8       textPaint.setAntiAlias(true);
9       textPaint.setColor(Color.BLACK);      //文本颜色为白色
10      textPaint.setTextSize(20);            //文字大小为 20
11      linePaint = new Paint();
12      linePaint.setColor(Color.BLUE);       //连接线的颜色默认为蓝色
13      linePaint.setAntiAlias(true);         //抗锯齿
```

```
14    linePaint.setStrokeWidth(2);              //直线宽度为 2 像素
15    linePaint.setStyle(Paint.Style.FILL);     //填充
16    numPaint = new Paint();
17    numPaint.setAntiAlias(true);
18    numPaint.setColor(Color.BLACK);           //文本颜色为白色
19    numPaint.setTextSize(18);                 //文字大小为 18
20    pointPaint = new Paint();
21    pointPaint.setAntiAlias(true);            //抗锯齿
22    pointPaint.setColor(Color.BLACK);         //黑色
23  }
```

接着开始绘制年份的坐标点,坐标点的本质其实是一个很小的圆形,这里我们需要使用一个 drawCircle(float cx,float cy,float radius,Paint paint)方法进行绘制,该方法的参数分别如下。

- cx:圆心的横坐标。
- cy:圆心的纵坐标。
- radius:圆的半径。
- paint:绘制圆要使用的画笔。

现在,是不是可以直接使用这个方法来绘制点呢?

事实上,我们现在只知道这些点需要绘制在 X 轴上,但并不知道它们绘制在哪个具体的位置。而且为了美观和清晰,我们希望每个点之间相隔的距离是一致的,同时最好平分坐标轴的空间(也就是当年份信息较多时,点之间的距离会变小,年份信息较少时,点之间的距离会变大)。

因此,需要在全局中设置一个变量 xSpace 来保存两个点之间 X 轴的距离,再定义一个数组 x[]保存每个点的横坐标。代码如下所示:

```
private int xSpace;              //两点之间 X 轴的间距
private int x[];                 //保存各个点的横坐标
```

先确定 x[]数组的长度。我们之前已经在集合中放入了模拟数据,而 x[]的长度其实就等于存放数据的集合的长度,因此直接进入 onDraw()方法,在方法中初始化数组 x[],代码如下所示:

```
x = new int[schoolRecruits.size()];          //X 轴坐标点的个数与具体的数据个数有关
```

然后,根据点的个数计算各个点之间 X 轴的间距 xSpace。我们希望 X 轴的左右各留一部分空间,然后剩下的空间由所有点平分,代码如下所示:

```
xSpace = (width - 200) / (x.length - 1);
//计算两个点之间的 X 轴间距,左右各留部分空间
```

接上步,还需要给每个点的横坐标赋值,代码如下所示:

```
for (int i = 0; i <x.length; i++) {          //初始化各个点的 X 轴坐标,起始点偏移 60 像素
    x[i] = 60 + i * xSpace;                  //左边偏移 60 个像素
```

}

小问题：如果直接在循环中使用 x[i]＝i＊xSpace 会发生什么事情？

现在我们已经为每个点的横坐标赋值完毕，同时也得出了每个点之间的距离，接下来我们开始绘制这些点，代码如下所示：

```
for(int i=0;i<x.length;i++){
    canvas.drawCircle(x[i],height-50,5,pointPaint);
}
```

现在回到视图页面，可以看到页面效果如图 4-52 所示。

提示：请前往 trendview.xml 布局文件中，给它添加一个背景色 #ccbbaa。

从上图中我们可以看出，X 坐标轴上出现了 3 个点，之所以是 3 个是因为我们设置的模拟数据中总共有 3 个年份的数据。点的数量是动态的，只要你在 setSchoolRecruit 方法中不断地添加不同年份的数据，点的数量就会自动增加。此时我们不需要再去修改 onDraw()方法中的任何代码，这样的写法大大地增加了代码的可读性和可用性。

接下来绘制年份的文字，年份文字位于 X 轴上年份点的正下方。我们需要将起始年份放在最左边，然后让它逐步递增，因此需要先在全局中定义一个变量来存储起始年份。

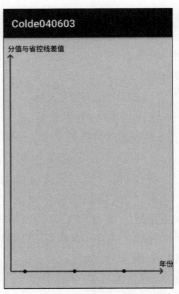

图 4-52　绘制 X 轴左边的点

```
private int year;                    //起始年份
```

然后在 onDraw()方法中获取数据中的起始年份，代码如下所示：

```
year = schoolRecruits.get(schoolRecruits.size() - 1).getsRecruitYear();
                                //获取起始年份
```

这段代码可以获取集合 schoolRecruits 中下标为"长度－1"的那个 SchoolRecruit 对象下的年份数据。现在我们可以再次使用 drawText()方法绘制年份文字信息。注意 drawText 方法要放在刚刚绘制年份信息点的循环内，代码如下所示：

```
for (int i=0;i<x.length;i++){
    canvas.drawText((year + i) + "", x[i] - 20, height - 20, numPaint);
                                //绘制年份文字
    canvas.drawCircle(x[i],height-50,5,pointPaint);    //绘制 X 轴上年份信息点
}
```

提示：之所以不使用 schoolRecruits.get(0).getsRecruitYear()来获取起始年份，是因为我们前面在往集合中插入 SchoolRecruit 对象时是按照 sc3～sc1 的顺序插入的，也

就意味着在集合中年份的排列顺序是 2015—2013，呈现一个倒序的排列。之所以这样设置，是为了契合"豹考通"教学版中的代码内容，因为"豹考通"教学版从"录取线查询"页面获取的数据也是倒序排列的。

此时绘制 X 轴信息的 TrendView.java 的所有代码如程序清单 4-65 所示。

程序清单 4-65：Code040601\app\src\main\java\cn\jxufe\iet\View\TrendView.java

```
1    public class TrendView extends View {
2      public TrendView(Context context) {
3        super(context);
4      }
5      public TrendView(Context context, AttributeSet attrs) {
6        super(context, attrs);
7      }
8      private Paint xyPaint;                        //定义一个坐标轴画笔
9      private Paint textPaint;                      //定义一个文字画笔
10     private Paint linePaint;                      //定义一个线条画笔
11     private Paint pointPaint,numPaint;            //定义点画笔和年份字体画笔
12     private int width,height;
13     private int arrawSpace = 8;                   //箭头的间距
14     private List<SchoolRecruit>schoolRecruits;
15     private int x[];                              //保存各个点的横坐标
16     private int xSpace;                           //保存各个点之间的横坐标距离
17     private int year;                             //起始年份
18     public List<SchoolRecruit>setSchoolRecruits(){
                                                     //建立集合,向集合中填充模拟数据
19     …//代码省略
20       return schoolRecruits;
21     }
22     public void init(){                           //该方法用于初始化画笔设置
23       xyPaint=new Paint();
24       xyPaint.setAntiAlias(true);
25       xyPaint.setStrokeWidth(3);
26       xyPaint.setColor(Color.RED);
27       //文本画笔的初始化
28       textPaint = new Paint();
29       textPaint.setAntiAlias(true);
30       textPaint.setColor(Color.BLACK);            //文本颜色为白色
31       textPaint.setTextSize(20);                  //文字大小为 20
32       linePaint = new Paint();
33       linePaint.setColor(Color.BLUE);             //连接线的颜色默认为蓝色
34       linePaint.setAntiAlias(true);               //抗锯齿
35       linePaint.setStrokeWidth(2);                //直线宽度为 2 像素
36       linePaint.setStyle(Paint.Style.FILL);       //填充
```

```java
37    numPaint = new Paint();
38    numPaint.setAntiAlias(true);
39    numPaint.setColor(Color.BLACK);              //文本颜色为白色
40    numPaint.setTextSize(18);                    //文字大小为18
41    pointPaint = new Paint();
42    pointPaint.setAntiAlias(true);               //抗锯齿
43    pointPaint.setColor(Color.BLACK);            //黑色
44    }
45    @Override
46    protected void onDraw(Canvas canvas) {
47        super.onDraw(canvas);
48        init();                                  //调用初始画笔的方法
49        width=this.getWidth();
50        height=this.getHeight();
51        canvas.drawLine(20, height - 50, width - 30, height - 50, xyPaint);
                                                   //绘制X轴
52        canvas.drawLine(20,height-50,20,50,xyPaint);
53        canvas.drawLine(width - 30 - arrawSpace, height - 50 - arrawSpace,
54            width - 30, height - 50, xyPaint);   //绘制X轴箭头
55        canvas.drawLine(width - 30 - arrawSpace, height - 50 + arrawSpace,
56            width - 30, height - 50, xyPaint);   //绘制X轴箭头
57        canvas.drawLine(20 + arrawSpace, 50 + arrawSpace, 20, 50, xyPaint);
                                                   //绘制Y轴箭头
58        canvas.drawLine(20 - arrawSpace, 50 + arrawSpace, 20, 50, xyPaint);
                                                   //绘制Y轴箭头
59        canvas.drawText("年份", width - 40, height - 50 - arrawSpace - 5,
60            textPaint);                          //X轴坐标提示文本
61        canvas.drawText("分值与省控线差值", 5 + arrawSpace, 40, textPaint);
                                                   //Y轴坐标提示文本
62        schoolRecruits=new ArrayList<SchoolRecruit>();
                                                   //初始化集合
63        schoolRecruits=setSchoolRecruits();      //向集合中填充数据
64        x=new int[schoolRecruits.size()];        //初始化数组,将数组的大小设置为集
                                                   //合内数据的长度
65        xSpace=(width-200)/(x.length-1);         //设置每个点之间X轴的距离
66        for (int i = 0; i<x.length; i++) {       //初始化各个点的X轴坐标,起始点偏移60像素
67            x[i] = 60 + i * xSpace;              //左边偏移40个像素
68        }
        year = schoolRecruits.get(schoolRecruits.size()-1).getsRecruitYear();
                                                   //获取起始年份
69        for (int i=0;i<x.length;i++){
70            canvas.drawText((year + i) + "", x[i] - 20, height - 20, numPaint);
                                                   //绘制年份文字
```

```
71     canvas.drawCircle(x[i],height-50,5,pointPaint);
                                              //绘制 X 轴上年份信息点
72    }
73   }
74  }
```

这时视图页面如图 4-53 所示,年份已经成功地绘制出来。

图 4-53　绘制 X 坐标轴上点的文字信息

3. 根据源数据绘制趋势图

现在我们根据集合中存储的数据绘制坐标点,在绘制坐标点之前要先说明"豹考通"教学版中趋势图的绘制依据。趋势图的横坐标代表"年份",纵坐标代表"分值与省控线差值"。那么这个"分值"指的是什么呢?

"豹考通"教学版的趋势图总共有三种不同类型的折线,分别有录取最高分、录取平均分、录取投档线三个指标参与其中。我们会分别使用这三个指标与省控线进行差值的计算并将它们的差值作为点的纵坐标。现在,我们编写一个 initDatas()方法并在这个方法中计算这些差值(见程序清单 4-66)。

需要先在全局中定义三个集合,分别用来保存录取最高分、录取平均分、投档线的数据,代码如下所示:

```
private List<Integer>firstDatas;        //录取最高分
private List<Integer>secondDatas;       //录取平均分
private List<Integer>thirdDatas;        //投档线
```

此外,还需要想到,有可能会存在某一年数据缺失的现象。这种情况下集合中是没有数据的,但如果不填充任何数据,程序会出现异常,因此我们定义一个变量来表示无数据时的情况(在全局中定义)。

private int NO_DATA=-1024;

其中-1024没有任何含义,只是用来填充集合,使集合不为空。

程序清单 4-66：TrendView.java 中的 initDatas() 方法

```
<!--初始化数据,获取分数与省控线之间的差距,如果不存在,则让其为 NO_DATA--!>
1   public void initDatas() {
2   firstDatas = new ArrayList<Integer>();
3   secondDatas = new ArrayList<Integer>();
4   thirdDatas = new ArrayList<Integer>();
5   for (int i = schoolRecruits.size() - 1; i >= 0; i--) {
6   int sRecruitHScore = schoolRecruits.get(i).getsRecruitHScore();
                                                                    //录取最高分
7   int controlLine = schoolRecruits.get(i).getControlLine();    //省控线
8   int enterLine = schoolRecruits.get(i).getEnterLine();        //投档线
9   int sRecruitAScore = schoolRecruits.get(i).getsRecruitAScore();
                                                                    //录取平均分
10  if (controlLine <= 0) {
11  firstDatas.add(NO_DATA);
12  secondDatas.add(NO_DATA);
13  thirdDatas.add(NO_DATA);
14  } else {
15  if (sRecruitHScore <= 0) {                    //如果录取最高分不存在
16  firstDatas.add(NO_DATA);
17  } else {
18  firstDatas.add(sRecruitHScore - controlLine);
                                                   //使用最高分减去省控线,计算差值
19  }
20  if (sRecruitAScore <= 0) {                    //如果录取平均分不存在
21  secondDatas.add(NO_DATA);
22  } else {
23  secondDatas.add(sRecruitAScore - controlLine);
                                                   //使用平均分减去省控线,计算差值
24  }
25  if (enterLine <= 0) {                         //如果投档线不存在
26  thirdDatas.add(NO_DATA);
27  } else {
28  thirdDatas.add(enterLine - controlLine);      //使用录取线减去省控线,计算差值
29  }
30  }
31  }
32  List<Integer> datas = new ArrayList<Integer>();
```

```
33    datas.addAll(firstDatas);
34    datas.addAll(secondDatas);
35    datas.addAll(thirdDatas);
```

通过以上代码,已经成功地把录取最高分、录取平均分、录取投档线这三个指标与省控线相减并存储到对应的集合(firstDatas、secondDatas 和 thirdDatas)中,最后再将这三个集合统一汇集到一个大集合 datas 中。

由于分数线的差值时高时低,因此很可能会存在由于差值过大(或者过小)导致趋势图超出屏幕或者分布太紧等错误。为此,我们需要为趋势线的 Y 轴设立一个中值,让趋势图中点的 Y 坐标都围绕这个中值排布,这样整个趋势图就会绘制在屏幕的中间,不至于太高也不至于太低。我们要找出这些差值中的最大值和最小值,然后使用最大值和最小值算出一个平均值。后面在绘制点并确定点的位置时,就使用这个平均值作为参考。

因此,需要在全局中定义如下三个变量:

```
private int maxData;            //最大值
private int minData;            //最小值
private int mindData;           //中间参考值
```

为了计算出一个集合中的最大值和最小值,我们自己编写两个方法,首先是获取最大值的方法 getMaxData(见程序清单 4-67)。

程序清单 4-67:Code040603\app\src\main\java\cn\jxufe\iet\TrendView.java

```
1    public int getMaxData(List<Integer> datas) {    //获取这组数据中最大的数据
2        int max = datas.get(0);                     //默认让第一个数据为最大数
3        for (int i = 0; i < datas.size(); i++) {
4            if (max < datas.get(i)) {              //如果有数据比最大数大,则将保留该数
5                max = datas.get(i);
6            }
7        }
8        return max;}
9
```

然后是获取最小值的方法 getMinData(见程序清单 4-68)。

程序清单 4-68:Code040603\app\src\main\java\cn\jxufe\iet\TrendView.java

```
1    public int getMinData(List<Integer> datas) {   //获取这组数据中最小的数据
2        int min = 1024;                             //为最大值
3        for (int i = 0; i < datas.size(); i++) {
4            int temp = datas.get(i);                //获取每一个数据
5            if (temp != -1024&& temp < min) {
6                min = temp;
7            }
8        }
```

```
9       return min;
10   }
```

接着在 initData() 方法中调用它们并求出平均值,此时 initData() 的代码如程序清单 4-69 所示。

程序清单 4-69:TrendView.java 中的 initData() 方法

```
1    public void initDatas() {
2        firstDatas = new ArrayList<Integer>();
3        secondDatas = new ArrayList<Integer>();
4        thirdDatas = new ArrayList<Integer>();
5        for (int i = schoolRecruits.size() - 1; i >= 0; i--) {
6            int sRecruitHScore = schoolRecruits.get(i).getsRecruitHScore();
                                                                        //录取最高分
7            int controlLine = schoolRecruits.get(i).getControlLine();    //省控线
8            int enterLine = schoolRecruits.get(i).getEnterLine();        //投档线
9            int sRecruitAScore = schoolRecruits.get(i).getsRecruitAScore();
                                                                        //录取平均分
10           if (controlLine <= 0) {
11               firstDatas.add(NO_DATA);
12               secondDatas.add(NO_DATA);
13               thirdDatas.add(NO_DATA);
14           } else {
15               if (sRecruitHScore <= 0) {                    //如果录取最高分不存在
16                   firstDatas.add(NO_DATA);
17               } else {
18                   firstDatas.add(sRecruitHScore - controlLine);
                                                    //使用最高分减去省控线,计算差值
19               }
20               if (sRecruitAScore <= 0) {                    //如果录取平均分不存在
21                   secondDatas.add(NO_DATA);
22               } else {
23                   secondDatas.add(sRecruitAScore - controlLine);
                                                    //使用平均分减去省控线,计算差值
24               }
25               if (enterLine <= 0) {                         //如果投档线不存在
26                   thirdDatas.add(NO_DATA);
27               } else {
28                   thirdDatas.add(enterLine - controlLine);  //使用录取线减去省控线,计算差值
29               }
30           }
31       }
32       List<Integer> datas = new ArrayList<Integer>();
33       datas.addAll(firstDatas);
```

```
34    datas.addAll(secondDatas);
35    datas.addAll(thirdDatas);
36    maxData = getMaxData(datas);              //获取所有数据中的最大值
37    minData = getMinData(datas);              //获取所有数据中的最小值
38    mindData = (maxData + minData) / 2;       //最大值与最小值的平均值,即中间值
```

数据和参考坐标已经准备完毕,接下来可以开始绘制趋势图。由于绘制的过程较为复杂,系统自带的 drawLine() 方法并不能满足绘制的要求,因此我们在"豹考通"教学版中重写了一个 drawLine() 方法,代码如程序清单 4-70 所示。

程序清单 4-70:TrendView.java 中的 drawLine() 方法

```
1   public void drawLine(List<Integer> datas, Canvas canvas, Paint linePaint,
2   boolean isTextUp) {
3       for (int i = 0; i < datas.size(); i++) {        //依次获取每一个数据
4           int data = datas.get(i);
5           if (data !=NO_DATA) {                        //如果对应的值有数据,则绘制
                //该数据相对于中间数据的偏移量,然后根据偏移量计算它的位置,偏移 1 对应 ySpace 个
                //像素,如果为负值表示在中线下方,如果为正值表示在中线上方
6               float point = (-(data - mindData)) * ySpace;   //该点相对于中线的纵坐标
7               canvas.drawCircle(x[i], centerHeight + point, 5,pointPaint);//绘制坐标点
8               if (isTextUp) {                          //isTextUp用来表示文字是在折线的上方还是下方
9                   canvas.drawText(data + "", x[i] - (data + "").length() * 6,
10                  centerHeight + point - textHeight / 2, numPaint);
                                                         //在上方绘制该点对应的文字信息
11              } else {
12                  canvas.drawText(data + "", x[i] - (data + "").length() * 6,
13                  centerHeight + point + textHeight, numPaint);
                                                         //在下方绘制该点对应的文字信息
14              }
15              if (i != (datas.size() - 1)
16                  && datas.get(i + 1) !=NO_DATA) {
                        //如果该点不是最后一个点,则需要绘制该点到下一个点的连线
17                  int nextData = (Integer) datas.get(i + 1);    //获取下一个数据
18                  float pointNext = (-(nextData - mindData)) * ySpace;
                                                         //该数据相对于中值的偏移量
19                  canvas.drawLine(x[i], centerHeight + point, x[i + 1],
20                  centerHeight + pointNext, linePaint);
21              }
22          }
23      }
24  }
```

此时代码会报错,因此我们还要再增加几个全局变量,如下所示:

private int ySpace = 2;   //纵向 1 个单位的间距,默认为 2

```
private int centerHeight;                    //中间参考值对应的纵坐标
private float textHeight;                    //文字高度
```

所有方法都写完了，只要在 onDraw() 方法中完成最后一步即可(见程序清单 4-71)。

程序清单 4-71：TrendView.java 中的 onDraw() 方法

```
1   protected void onDraw(Canvas canvas) {
2       super.onDraw(canvas);
3       init();                                                //调用初始画笔的方法
4       width=this.getWidth();
5       height=this.getHeight();
6       canvas.drawLine(20, height - 50, width - 30, height - 50, xyPaint);
                                                               //绘制 X 轴
7       canvas.drawLine(20,height-50,20,50,xyPaint);
8       canvas.drawLine(width - 30 - arrawSpace, height - 50 - arrawSpace,
9           width - 30, height - 50, xyPaint);                 //绘制 X 轴箭头
10      canvas.drawLine(width - 30 - arrawSpace, height - 50 + arrawSpace,
11          width - 30, height - 50, xyPaint);                 //绘制 X 轴箭头
12      canvas.drawLine(20 + arrawSpace, 50 + arrawSpace, 20, 50, xyPaint);
                                                               //绘制 Y 轴箭头
13      canvas.drawLine(20 - arrawSpace, 50 + arrawSpace, 20, 50, xyPaint);
                                                               //绘制 Y 轴箭头
14      canvas.drawText("年份", width - 40, height - 50 - arrawSpace - 5,
15          textPaint);                                        //X 轴坐标提示文本
16      canvas.drawText("分值与省控线差值", 5 + arrawSpace, 40, textPaint);
                                                               //Y 轴坐标提示文本
17      schoolRecruits=new ArrayList<SchoolRecruit>();         //初始化集合
18      schoolRecruits=setSchoolRecruits();                    //向集合中填充数据
19      x=new int[schoolRecruits.size()];
20      xSpace=(width-200)/(x.length-1);
21      for (int i = 0; i < x.length; i++) {
                                          //初始化各个点的 X 轴坐标,起始点偏移 60 像素
22          x[i] = 60 + i * xSpace;                            //左边偏移 40 个像素
23      }
24      year = schoolRecruits.get(schoolRecruits.size()-1).getsRecruitYear();
                                                               //获取起始年份
25      for (int i=0;i<x.length;i++){
26          canvas.drawText((year + i) + "", x[i] - 20, height - 20, numPaint);
27          canvas.drawCircle(x[i],height-50,5,pointPaint);
28      }
```

```
29    initDatas();                                          //调用计算差值和平均值的方法
30    Paint.FontMetrics fontMetrics = textPaint.getFontMetrics();
                                                            //获取文字坐标
31    textHeight = fontMetrics.bottom - fontMetrics.top;
                                                            //文字底部坐标减去顶部坐标
32    centerHeight = height / 2 - 10;
             //获取控件的 Y 轴中线,中线为中间数据,比该数据大的在上方,比该数据小的在下方
33    if (firstDatas != null) {                             //绘制录取最高分
34        linePaint.setColor(Color.RED);
35        canvas.drawText("录取最高分", width - 105, 20, linePaint);
                                                            //绘制右上角提示文字
36        canvas.drawLine(width - 130, 15, width - 110, 15, linePaint);
                                                            //绘制右上角提示线条
37        drawLine(firstDatas, canvas, linePaint, true);    //绘制趋势线
38    }
39    if (secondDatas != null) {                            //绘制录取平均分
40        linePaint.setColor(Color.BLUE);
41        canvas.drawText("录取平均分", width - 105, 45, linePaint);
                                                            //绘制右上角提示文字
42        canvas.drawLine(width - 130, 40, width - 110, 40, linePaint);
                                                            //绘制右上角提示线条
44        drawLine(secondDatas, canvas, linePaint, true);   //绘制趋势线
44    }
45    if (thirdDatas != null) {                             //绘制投档线
46        linePaint.setColor(Color.YELLOW);
47        canvas.drawText("录取投档线", width - 105, 70, linePaint);
                                                            //绘制右上角提示文字
48        canvas.drawLine(width - 130, 65, width - 110, 65, linePaint);
                                                            //绘制右上角提示线条
49        drawLine(thirdDatas, canvas, linePaint, true);    //绘制趋势线
50    }
51 }
```

这时回到视图界面,可观察到图 4-54 所示的效果。

至此,我们已经成功地使用自定义的集合数据完成了趋势图的绘制。

4.6.4 用绘图类绘制趋势线

1. 基本实现过程

我们使用来自录取线页面的数据绘制趋势图。录取线页面是在前几章中已经完成的模块,为了节约时间,可以直接导入项目 Code040603,在 Code040603 的基础上完成本节要讲的内容。

选择 File→Open 选项,找到项目 Code040603 直接打开即可,如图 4-55 所示。

项目结构如图 4-56 所示。

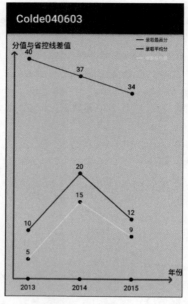

图 4-54 完成趋势图的绘制

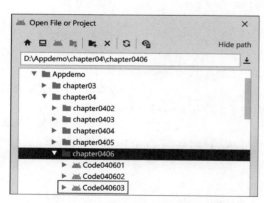

图 4-55 打开项目 Code040603

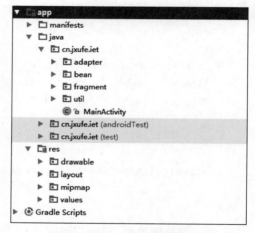

图 4-56 Code040603 项目结构

导入成功后,将之前已经做好的趋势图类和对应的布局文件复制进来,注意不要忘记建包。此时项目结构如图 4-57 所示。

我们先进入"录取线查询"的 Java 逻辑代码模块 EnrollScoreFragment 中。它的路径为 cn\jxufe\iet\fragment\EnrollScoreFragment.java。可以看到图 4-58 中被框起来的代码里已经定义好一个趋势图按钮,在"豹考通"教学版中,用户通过单击这个按钮进入趋势图页面,但此时我们并没有为这个按钮添加任何逻辑。

根据前面的知识,我们为这个按钮设立一个单击事件,找到 onCreateView() 方法,在其中加入如下代码。

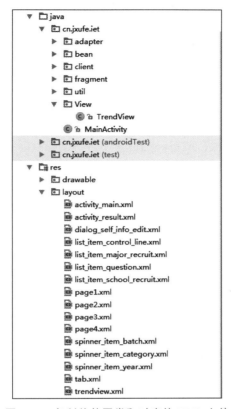

图 4-57　复制趋势图类和对应的 XML 文件

```
public class EnrollScoreFragment extends Fragment {
    private Spinner categorySpinner, batchSpinner, sourceAre
    private RadioButton isSearchByMajor;// 是否按专业查询
    private Button searchBtn;// 查询按钮
    private Button trendBtn;// 生成趋势图按钮
    private TextView titleView;// 显示标题
```

图 4-58　趋势图按钮

```
trendBtn.setOnClickListener(new View.OnClickListener() {
    //生成趋势图的按钮的事件处理
@Override
public void onClick(View v) {
        goToShowTrend();                                    //显示趋势图
    }
});
```

goToShowTrend()方法是我们自己定义的一个方法，接下来就去编写它。首先，我们需要知道这个方法中到底需要进行一些什么操作。

① 获取当前页面中的数据。

② 获取完数据后,将这些数据发送到趋势图页面中。
③ 让页面跳转到趋势图页面。

操作①的实现代码如程序清单 4-72 所示。

程序清单 4-72:EnrollScoreFragment.java 中的 goToShowTrend() 方法

```
1   public void goToShowTrend() {//跳转到显示趋势图的页面
2   Bundle bundle = new Bundle();
3       String schoolRecruitsResult = gson.toJson(schoolRecruits);
4       bundle.putString("schoolRecruitsResult", schoolRecruitsResult);
                                                    //学校录取线结果信息
5   bundle.putString("schoolName", schoolName);     //学校名称
6   bundle.putString("batchName",batches[batchId]); //批次信息
7   bundle.putString("categoryName",categories[categoryId]);
                                                    //类别信息
8   bundle.putString("studentAreaName", provinces.get(sourceAreaId));
                                                    //生源地
9   }
```

第 5 行代码报错,因为需要在全局中加入 schoolName 变量。

private String schoolName;   //学校名称,标题上需要显示

现在我们先来看 Bundle 类。Bundle 类用于传递数据,它保存的数据是以键-值对的形式存在的。两个 Activity 之间的通信可以通过 Bundle 来实现,也就是说,我们可以将一个 Activity 中的数据存储在 Bundle 类中,然后将这个 Bundle 传给另一个 Activity。在录取线查询页面中取得录取线结果、学校名称、批次信息等数据,然后使用 putString 方法将它们封装在 Bundle 对象中。

操作②与操作③的实现代码见下一节。

2. 使用 Android Studio 导入 Gson 包

我们还注意到代码 String schoolRecruitsResult = gson.toJson(schoolRecruits);,这行代码此时报错,因为 Gson 这个对象还没有定义。我们先在全局中加入下列代码:

private Gson gson = new Gson();

但代码仍然报错,这是因为 Gson 并不是 Android 自带的 API,我们需要通过外部导入来引入这个 Gson 类。操作方法如下所示:

① 选择 File→Project Structure 选项。
② 在弹出的窗口中,选中 Dependencies→app 并单击"＋"按钮。
③ 选择 Library Dependency。
④ 在搜索框中输入 gson。
⑤ 经过一段时间的等待后,可以看到搜索结果,我们选择以 com.google.code.gson:开头的两个文件并单击 OK 按钮。
⑥ 在经过一段等待(实际上是在下载)时间后,我们可以发现 Gson 包已经成功地导

入项目中。

现在我们可以使用 Alt＋Enter 键来导入 Gson 包。

Gson 是一种 Google 开发的 Java API,可用于转换 Java 对象和 JSON 对象。回到之前的 goToShowTrend（）方法中,我们使用了 String schoolRecruitsResult ＝ gson.toJson(schoolRecruits);将学校的录取线集合转换为一串 JSON 数组。

实现获取数据、将获取的数据发送到趋势图页面并跳转到趋势图页面的关键代码如程序清单 4-73 所示。

程序清单 4-73：Code040604\app\src\main\java\cn\jxufe\iet\fragment\EnrollscoreFragment.java

```
1    Intent intent = new Intent(getActivity(), ResultActivity.class);
                                                    //设置跳转位置
2    intent.putExtra("bundle", bundle);             //传递数据
3    intent.putExtra("fragment", "SchoolRecruitTrendFragment");
4    startActivity(intent);                         //跳转页面
```

intent 的中文意思是"意图",Android 中提供了 Intent 机制来协助应用间的交互与通信。Intent 负责对应用中一次操作的动作、动作涉及的数据以及附加数据进行描述,Android 则根据此 Intent 的描述负责找到对应的组件,将 Intent 传递给调用的组件并完成组件的调用。

此时,goToShowTrend()方法的代码如程序清单 4-74 所示。

程序清单 4-74：Code040604\app\src\main\java\cn\jxufe\iet\View\TrendView.java

```
1    public void goToShowTrend() {                   //跳转到显示趋势图的页面
2        Bundle bundle = new Bundle();
3        String schoolRecruitsResult = gson.toJson(schoolRecruits);
4        bundle.putString("schoolRecruitsResult", schoolRecruitsResult);
                                                    //学校录取线结果信息
5        bundle.putString("schoolName", schoolName);                //学校名称
6        bundle.putString("batchName",batches[batchId]);            //批次信息
7        bundle.putString("categoryName",categories[categoryId]);   //类别信息
8        bundle.putString("studentAreaName", provinces.get(sourceAreaId));
                                                                   //生源地
9        Intent intent = new Intent(getActivity(), ResultActivity.class);
10       intent.putExtra("bundle", bundle);                         //传递数据
11       intent.putExtra("fragment", "SchoolRecruitTrendFragment");
12       startActivity(intent);                                     //跳转页面
13   }
```

4.6.5　编写趋势线相关 Activity 和 Fragment

接下来我们在 cn.jxufe.iet 文件夹中建立一个 client 包并在包中建立一个 Java 类,命名为 ResultActivity.java,如图 4-59 所示。

这个 ResultActivity.java 是我们建立的一个新 Activity，这个 Activity 可以接收到 SchoolRecruitTrendFragment 发送的 Bundle 数据，并且让页面跳转到趋势图页面。这个 ResultActivity.java 是一个完整的 Activity，它需要继承 FragmentActivity。然后我们再建立一个 SchoolRecruitTrendFragment.java 类来编写趋势图页面的业务逻辑并让这个类继承 Fragment。也就是说，SchoolRecruitTrendFragment 是位于 ResultActivity 中的 Fragment，如图 4-60 所示。

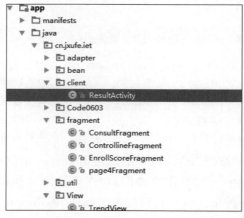

图 4-59　创建 ResultActivity 类

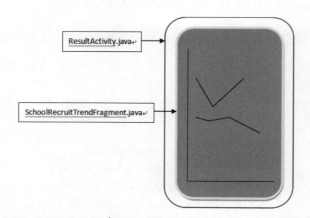

图 4-60　ResultActivity 与 SchoolRecruitTrendFragment 之间的关系

1. 编写趋势图的 Fragment 类

我们暂且不编写 ResultActivity.java 的代码，先完成 SchoolRecruitTrendFragment.java 的编写，找到 fragment 文件夹，在该文件夹下建立一个名为 SchoolRecruitTrendFragment 的 Java 类。

SchoolRecruitTrendFragment.java 的代码如程序清单 4-75 所示。

程序清单 4-75：Code040604\app\src\main\java\cn\jxufe\iet\fragment\SchoolRecruitTrendFragment.java

```
1    public class SchoolRecruitTrendFragment extends Fragment {
```

```
2    private TextView titleView;                              //显示标题的文本显示框
3    private Gson gson = new Gson();
4    private String schoolName;                               //学校名称
5    private String result;                                   //录取线查询结果
6    private String batchName;                                //批次
7    private String categoryName;                             //类别
8    private String studentAreaName;                          //生源地
9    private List<SchoolRecruit>schoolRecruits;
10   private TrendView trendView;                             //自定义趋势图控件
11   public View onCreateView (LayoutInflater inflater, @ Nullable ViewGroup
     container, @Nullable Bundle savedInstanceState) {
12   View view = inflater.inflate(R.layout.trendview, container,false);
13   trendView=(TrendView)view.findViewById(R.id.trend);      //获取自定义趋势图控件
14   trendView.setSchoolRecruits(schoolRecruits);
15   return view;
16   }
17   public void onCreate(Bundle savedInstanceState) {
18   super.onCreate(savedInstanceState);
19   Bundle bundle = getArguments();                          //获取传递过来的参数
20   if (bundle != null) {
21   schoolName = bundle.getString("schoolName");             //从参数中获取学校名称
22   result = bundle.getString("schoolRecruitsResult");       //从参数中获取查询结果
23   batchName = bundle.getString("batchName");               //获取批次
24   categoryName = bundle.getString("categoryName");         //获取类别
25   studentAreaName = bundle.getString("studentAreaName");   //获取生源地
26   schoolRecruits = gson.fromJson(result,
27   new TypeToken<List<SchoolRecruit>>() {}.getType());      //获取学校录取线信息
28   }
29   }
30   }
```

当我们敲完代码后发现第 14 行代码 trendView.setSchoolRecruits(schoolRecruits) 报错,这是因为在我们的趋势图类 TrendView.java 中还没有将 setSchoolRecruits() 方法进行更改。现在我们进入 TrendView 类,找到 setSchoolRecruits() 方法,将它的代码改为如程序清单 4-76 所示。

程序清单 4-76:Code040604\app\src\main\java\cn\jxufe\iet\View\TrendView.java

```
1  public void setSchoolRecruits(List<SchoolRecruit> schoolRecruits) {
2  this.schoolRecruits = schoolRecruits;
3  init();
4  isDrawLine=true;                           //初始化数据后,即可绘制折线图
5  invalidate();                              //重新绘制,刷新显示
6  }
```

除此之外还要在这个类的全局中加入一个 boolean 类型的变量 isDrawLine。

private boolean isDrawLine = false;

现在再回到 SchoolRecruitTrendFragment.java 中,发现代码已经不会报错,但这个时候 TrendView.java 的代码结构也需要发生一些改变。代码如程序清单 4-77 所示。

程序清单 4-77:Code040604\app\src\main\java\cn\jxufe\iet\View\TrendView.java

```
1    public class TrendView extends View {
2    publicTrendView(Context context) {
3    super(context);
4        }
5    publicTrendView(Context context, AttributeSet attrs) {
6    super(context, attrs);
7        }
8    private Paint xyPaint;                          //定义一个坐标轴画笔
9        ……
10   public void setSchoolRecruits(List<SchoolRecruit> schoolRecruits) {
11   this.schoolRecruits = schoolRecruits;
12       init();                                    //调用初始化画笔方法
13   isDrawLine=true;                               //初始化数据后,即可绘制折线图
14   invalidate();                                  //重新绘制,刷新显示
15   }
16   public void init(){                            //该方法用于初始化画笔设置
17   ……
18   x=new int[schoolRecruits.size()];
19   year = schoolRecruits.get(schoolRecruits.size()-1).getsRecruitYear();
                                                    //获取起始年份
20   Paint.FontMetrics fontMetrics = textPaint.getFontMetrics();
21   textHeight = fontMetrics.bottom - fontMetrics.top;
                                                    //文字底部坐标减去顶部坐标
22   initDatas();                                   //调用初始化数据方法
23       }
24   public int getMaxData(List<Integer> datas) {//获取这组数据中最大的数据
25           …………
26       }
27   public int getMinData(List<Integer> datas) {//获取这组数据中最小的数据
28   …………
29       }
30       @Override
31   protected void onDraw(Canvas canvas) {
32   super.onDraw(canvas);
33   width=this.getWidth();
34   height=this.getHeight();
```

```
35  xSpace=(width-200)/(x.length-1);
36  ySpace = (height - 160) / Math.abs(maxData - minData);    //上下各留 65
37  for (int i = 0; i < x.length; i++) {//初始化各个点的 X 轴坐标,起始点偏移 60 像素
38     x[i] = 60 + i * xSpace;                         //左边偏移 40 个像素
39  }
40  centerHeight = height / 2 - 10;
          //获取控件的 Y 轴中线,中线为中间数据,比该数据大的在上方,比该数据小的在下方
41  drawXY(canvas);
42  if(firstDatas != null) {                        //绘制录取最高分
44     linePaint.setColor(Color.RED);
44          canvas.drawText("录取最高分", width - 105, 20, linePaint);
45     canvas.drawLine(width - 130, 15, width - 110, 15, linePaint);
46     drawLine(firstDatas, canvas, linePaint, true);
47       }
48  if(secondDatas != null) {                       //绘制录取平均分
49     linePaint.setColor(Color.BLUE);
50          canvas.drawText("录取平均分", width - 105, 45, linePaint);
51     canvas.drawLine(width - 130, 40, width - 110, 40, linePaint);
52     drawLine(secondDatas, canvas, linePaint, true);
53       }
54  if(thirdDatas != null) {                        //绘制投档线
55     linePaint.setColor(Color.YELLOW);
56          canvas.drawText("录取投档线", width - 105, 70, linePaint);
57     canvas.drawLine(width - 130, 65, width - 110, 65, linePaint);
58     drawLine(thirdDatas, canvas, linePaint, true);
59       }
60     }
61  public void drawXY(Canvas canvas){
62  …………
63  }
64  <!--初始化数据,获取分数与省控线之间的差距,如果不存在,则让其为 NO_DATA--!>
65  public void initDatas() {
66  …………
67  }
68  public void drawLine(List<Integer> datas, Canvas canvas, Paint linePaint,
69     booleanisTextUp) {
70  …………
71       }
72  }
```

2. 编写趋势图的 Activity

下面我们编写与 ResultActivity.java 相对应的 XML 文件,找到 layout 文件夹,在里面建立一个名为 activity_result 的 XML 文件(见程序清单 4-78 所示)。

程序清单 4-78：Code040604\app\src\main\res\layout\activity_result.xml

```
1   <LinearLayout xmlns:android="http://schemas.android.com/apk/res/android"
2   android:layout_width="match_parent"
3   android:layout_height="match_parent"
4   android:background="#ccbbaa"
5   android:orientation="vertical">
6   <FrameLayout
7   android:id="@+id/resultContent"
8   android:layout_width="match_parent"
9   android:layout_height="match_parent"/>
10  </LinearLayout>
```

最后编写 ResultActivity.java，也就是趋势图页面的整个 Activity 的 Java 代码（见程序清单 4-79 所示）。

程序清单 4-79：Code040604\app\src\main\java\cn\jxufe\iet\client\ResultActivity.java

```
1   public class ResultActivity extends FragmentActivity {
2       private Intent intent;                              //Intent 对象
3       private Bundle bundle;                              //封装数据的 Bundle 对象
4       private String fragmentString;                      //要显示的 Fragment 的名称
5       @Override
6       protected void onCreate(Bundle savedInstanceState) {
7           super.onCreate(savedInstanceState);
8           requestWindowFeature(Window.FEATURE_NO_TITLE);  //去除标题栏
9           setContentView(R.layout.activity_result);
10          intent = getIntent();                           //获取传递的 Intent
11          fragmentString = intent.getStringExtra("fragment");
                                                            //获取传递的 Fragment 的名称
12          bundle = intent.getBundleExtra("bundle");       //获取 Intent 中的数据集
13          if (fragmentString.equals("SchoolRecruitTrendFragment")) {
                                                            //根据获取 Fragment 名称显示对应的 Fragment
14              SchoolRecruitTrendFragment enrollTrendFragment = new
    SchoolRecruitTrendFragment();
15              enrollTrendFragment.setArguments(bundle);
16              getSupportFragmentManager().beginTransaction()
17                  .replace(R.id.resultContent, enrollTrendFragment).commit();
18          }
19      }
20      public void goBack(View view) {                     //返回按钮的单击事件处理
21          this.finish();
22      }
23  }
```

3. 为 ResultActivity 进行注册

完成 ResultActivity 的代码编写后，还需要进入 AndroidManifest.xml 文件中注册这个 Activity，如程序清单 4-80 所示。

程序清单 4-80：Code040604\app\src\main\AndroidManifest.xml

```
1   <?xml version="1.0" encoding="utf-8"?>
2   <manifest xmlns:android="http://schemas.android.com/apk/res/android"
3       package="cn.jxufe.iet">
4       <application
5           android:allowBackup="true"
6           android:icon="@mipmap/ic_launcher"
7           android:label="@string/app_name"
8           android:supportsRtl="true"
9           android:theme="@style/AppTheme">
10          <activity android:name="cn.jxufe.iet.Code0603.MainActivity">
11              <intent-filter>
12                  <action android:name="android.intent.action.MAIN"/>
13                  <category android:name="android.intent.category.LAUNCHER"/>
14              </intent-filter>
15          </activity>
    <!--注册 ResultActivity--!>
16          <activity android:name="cn.jxufe.iet.client.ResultActivity"></activity>
17      </application>
18  </manifest>
```

现在我们运行程序，首先进入录取线查询页面查找出一系列录取线数据（见图 4-61），然后单击按钮"历年分数与省控线差值趋势图"，可以发现趋势图已经成功地呈现出来（见图 4-62）。

图 4-61　在录取线查询界面获得数据

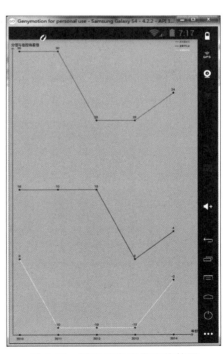

图 4-62　录取线查询界面趋势图

4.7 本章小结

"豹考通"教学版项目分为本地版和网络版,本章主要实现本地版,同时为后面章节实现网络版打下基础。"豹考通"项目分为"省控线查询""录取线查询""报考咨询""个人中心"四个模块。"省控线查询"和"录取线查询"两个模块比较类似,都是具有 Spinner、事件监听、列表等功能或控件,因此以实现"省控线查询"模块为例实现两个查询模块。"咨询模块"主要以自定义列表形式进行实现,其中包括获取时间、获取数据、对话框等功能,数据以本地类文件提供为主,在后面讲解用 SQLite 进行替换。"个人基本信息"模块中的个人高考信息数据保存在 SharedPreferences 生成文件中,用户点击"编辑"按钮后会弹出对话框供其编辑个人信息。最后,完成录取线的趋势图,先是实现绘图的自定义 View 类,使用页面跳转携带数据实现录取线趋势图。

4.8 课后练习

1. 继续完成未完成的任务,将选项卡中的文字内容修改为"省控线""录取线""报考咨询""个人中心"。

2. 用 Spinner 列表尝试另外一种实现方式,在 XML 文件中通过<string-array>进行指定,然后为 Spinner 控件指定 android:entries 属性。

3. 尝试编写一个登录的小案例,该页面应该有一个登录按钮和两个输入框。当输入账号 admin 和密码 123 时,可弹出一个 Toast 提示"登录成功",若账号或密码输入错误,则弹出一个 Toast 提示"登录失败"。登录成功后需要将账号和密码存储在 SharedPreferences 中,当用户第二次打开案例时,程序会从 SharedPreferences 中取出账号和密码,并将它们自动填写在输入框中。

4. 请设计并实现图 4-63 所示的界面效果。

(a)　　　　　　　(b)

图 4-63　界面效果图

界面要求：
- 最上面为一个图片控件。中间三行采用表格布局进行显示，其中 TextView 字体采用 22sp，第一行为 Spinner 控件，第二行和第三行为 RadioButton，最下面为按钮。背景颜色为♯8080C0，宽度 200dp，高度 40dp。
- Spinner 控件设置背景。
- 使用 shape 为根标签的 XML 文件为使用省份数组显示 Spinner 列表，如图 4-63（b）所示。

第5章

数据库开发

5.1 本章简介

在 Android 应用开发中经常用到的数据库有两个：一个是轻量级关系型数据库 SQLite，另一个是关系型数据库 MySQL。在 Android 本地经常使用和数据量不是太大的数据一般存放在 SQLite 数据中，因此作为 Android 程序员要学会开发 SQLite 数据库，其中包括 SQLite 特点、创建、操作、导出等。在"豹考通"本地教学版中，Spinner 中的数据使用 Java 类从文件中访问，不过这种方式并不常用，一般从 SQLite 中获取。因此，本章带领大家创建"豹考通"SQLite 数据库并导入数据，在项目中使用 SQLite 常用类获取操作数据并显示在 Spinner 中。在 Android 开发中，MySQL 数据库为服务器端提供数据，一般数据量要比 SQLite 大得多，同样我们要学会安装 MySQL 数据库、如何设计数据库、创建 MySQL 数据库、导入导出数据以及使用 Java 代码如何连接访问数据库。

5.2 SQLite

5.2.1 了解 SQLite

SQLite 是一款轻量级、跨平台的关系型数据库。既然号称关系型数据库，那它就支持多数 SQL92 标准（如视图、触发器、事务等），这里不准备一一细说。它的运算速度非常快，并且占用资源很少，通常只需要几百 KB 的内存就可以，因而特别适合在移动设备上使用。Android 系统内置了数据库，SQLite 是 D.Richard Hipp 用 C 语言编写的开源嵌入式数据库引擎。它支持大多数的 SQL92 标准，并且可以在所有主要的操作系统上运行。

SQLite 由以下几个部分组成：SQL 编译器、内核、后端以及附件。SQLite 通过利用虚拟机和虚拟数据库引擎（VDBE），使调试、修改和扩展 SQLite 的内核变得更加方便。所有 SQL 语句都被编译成易读的可在 SQLite 虚拟机中执行的程序集。SQLite 支持 NULL、INTEGER、REAL、TEXT 和 BLOB 数据类型，分别代表空值、整型值、浮点值、字符串文本和二进制对象。SQLite 不仅支持标准的 SQL 语法，还遵循数据库的 ACID 事务，因此只要你以前使用过其他的关系型数据库，就可以很快地上手 SQLite。而 SQLite 又比一般的数据库要简单得多，它甚至不用设置用户名和密码就可以使用。Android 正是把这个功能极为强大的数据库嵌入系统当中，才使得本地持久化的功能有了一次质的飞跃。SQLite 系统架构如图 5-1 所示。

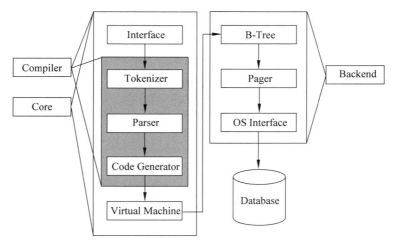

图 5-1　SQLite 系统架构

5.2.2　安装 SQLite

SQLite 的官方下载地址为 http://www.sqlite.org/download.html，上面提供了多种版本的 SQLite，如图 5-2 所示。

图 5-2　SQLite 官方下载网站

这里我们选择下载名为 sqlite-dll-win64-x64-3180000.zip 的版本，如图 5-3 所示。

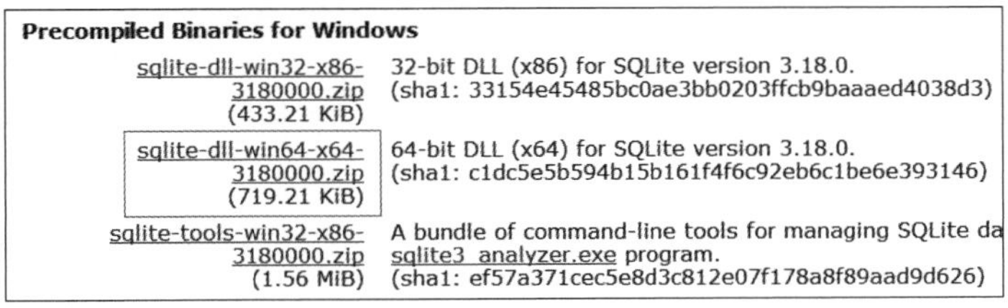

图 5-3　下载相对应的版本号

下载完后直接解压到磁盘上，可以看到解压后有两个文件（一个是 sqlite3.def，另一个是 sqlite3.dll），如图 5-4 所示。

图 5-4　解压到磁盘

接下来需要把 SQLite 加入 Path 环境变量中（加入环境变量是为了更加方便地使用 SQLite）。右键"我的电脑"，选择"属性"选项，再选择"高级系统设置"，如图 5-5 所示。

图 5-5　选择"高级系统设置"

然后选择"环境变量"按钮，如图 5-6 所示。

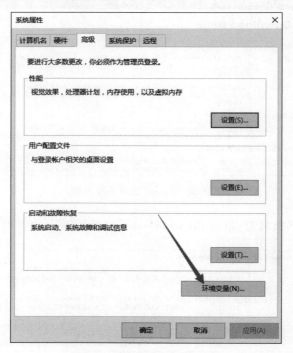

图 5-6　选择"环境变量"按钮

在"系统变量"框中找到 Path,双击进入编辑页面,如图 5-7 所示。

图 5-7　编辑 Path

将解压后的文件夹目录加到 Path 值后面(注意是文件夹目录,例如我本机的目录是 H:\sqlite-dll-win64-x64-3180000),如图 5-8 所示。

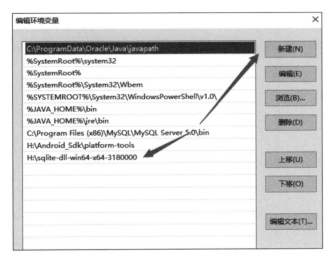

图 5-8　添加文件夹目录到 Path

然后打开"命令提示符"界面(Win+R 键),输入 sqlite3,如果弹出如图 5-9 所示的消息,就表示安装成功。

图 5-9　SQLite 安装成功

5.3 创建 SQLite 数据库

shell 脚本虽然提供了很强大的功能,但是使用起来不够方便,幸运的是,SQLite 有很多开源且优秀的 DBMS。本章我们将介绍如何使用管理工具保存本地数据并建立 SQLite 数据库,当你在项目中需要使用这些数据时,直接将 SQLite 数据库引用过来便可。这里我们将使用一款叫作 Navicat for SQLite 的软件,以它为例进行讲解,官网地址为 https://www.navicat.com/download/navicat-for-sqlite。这个软件是绿色免安装版,工具的下载和安装较简单,可以自行完成(解压直接运行即可),这里不多介绍。

5.3.1 创建 bkt 数据库

首先,打开 Navicat for SQLite 软件,出现图 5-10 所示的页面。

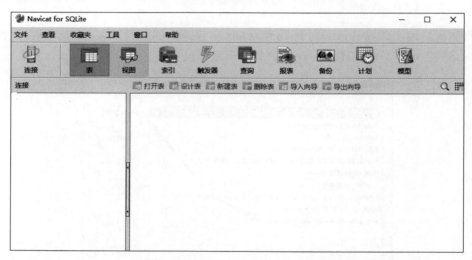

图 5-10　Navicat for SQLite 主要操作界面

然后,单击左上角"连接"按钮,出现"新建连接"界面。我们在"连接名"这一栏输入 bkt,类型选择"新建 SQLite 3"。需要注意的是,数据库文件一定不要写错或者不写,不然会报"新建连接出现 I/O error 105"错误。

"数据库文件"栏主要是提醒我们新建的 SQLite 3 文件需要保存在什么地方,我们选择桌面就行,然后单击"确定"按钮,如图 5-11 所示。

5.3.2 创建 area 表

之后在软件的左上方可以看到出现了 bkt 新连接。我们依次双击 bkt 及 main 目录,可看到在 main 目录下面出现了"表"选项,但是它目前还是空的。这里我们先创建 area 表。

右击"表",选择"新建表"选项,如图 5-12 所示。

在表的结构中输入"名""类型""长度"等属性,如图 5-13 所示。

图 5-11 新建 bkt 连接及数据库

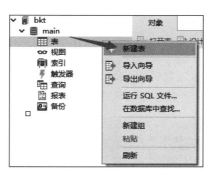

图 5-12 新建表

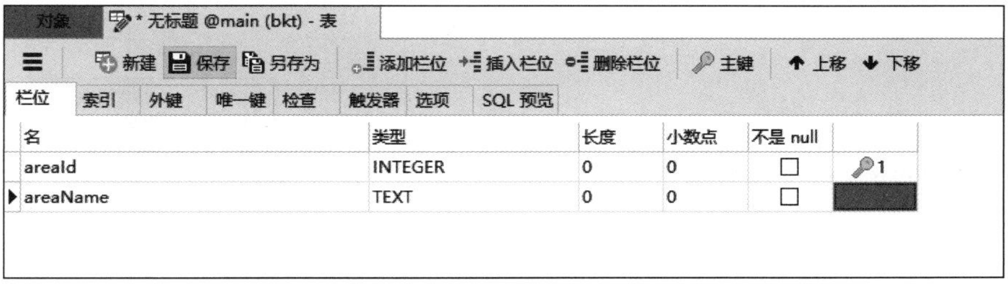

名	类型	长度	小数点	不是 null	
areaId	INTEGER	0	0	☐	🔑1
areaName	TEXT	0	0	☐	

图 5-13 创建 area 表的结构

单击"保存"按钮,输入表名 area,然后单击"确定"按钮,再在左边刷新表便可,如图 5-14 所示。

5.3.3 插入 area 表数据

建立了新表的结构及属性之后,我们在表中插入项目需要用到的数据。由于这里只有 4 条记录,因此可以选择直接在表中输入,然后保存便可,如图 5-15 所示。

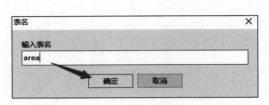

图 5-14 输入表名 area

图 5-15 直接输入数据

当然我们也可以使用 SQL 语句向表中插入数据。单击软件上方的"查询"按钮,然后单击"新建查询",如图 5-16 所示。

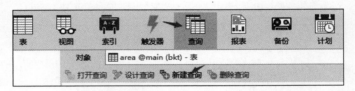

图 5-16 新建查询

在查询编辑器页面输入插入数据的 SQL 语句并依次执行,如图 5-17 所示。
然后刷新 area 表,可以看到数据已插入表中,如图 5-18 所示。

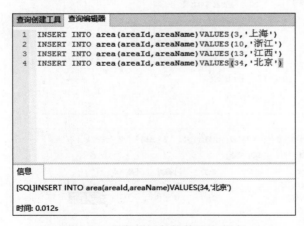

图 5-17 运行插入数据的 SQL 语句　　　图 5-18 使用 SQL 语句插入数据

5.3.4 创建 school 表

我们使用相同的方式创建 school 表,右击"表",选择"新建表"选项,如图 5-19 所示。

图 5-19 新建 school 表

然后根据数据分析得出的表结构依次建好各个属性("名""类型""长度"等),如图 5-20 所示。

图 5-20 建立 school 表结构

接着单击"保存"按钮,也可以使用快捷键 Ctrl+S 保存并输入表名 school,再单击"确定"按钮,如图 5-21 所示。

图 5-21 保存 school 表

5.3.5 导入 school 表数据

因为 school 表中存的数据有几百条,所以用直接输入或 SQL 语句插入的方式会显得很复杂,我们这里使用将 Excel 表中的数据导入 SQLite 数据库表的方法来导入数据。

先将需要用到的数据整理到 Excel 表中,部分记录如图 5-22 所示。

	A	B	C	D	E	F	G
1	schoolId	schoolName	schoolCode	areaId	is985	is211	isMinistry
2	5	北京大学	10001	34	1	1	1
3	6	清华大学	10003	34	1	1	1
4	8	浙江大学	10335	10	1	1	1
5	14	复旦大学	10246	3	1	1	1
6	16	上海中医药大学	10268	3	0	0	0
7	22	同济大学	10247	3	1	1	1
8	27	中国人民大学	10002	34	1	1	1
9	28	北京理工大学	10007	34	1	1	1
10	31	上海大学	10280	3	0	1	0
11	39	中央财经大学	10034	34	0	1	1
12	44	上海交通大学	10248	3	1	1	1
13	45	上海财经大学	10272	3	0	1	1
14	48	南昌大学	10403	13	0	1	0
15	51	北京师范大学	10027	34	1	1	1
16	57	北京航空航天大学	10006	34	1	1	1
17	59	华东师范大学	10269	3	1	1	1
18	61	北京邮电大学	10013	34	0	1	1
19	68	华东理工大学	10251	3	0	1	1
20	75	中国传媒大学	10033	34	0	1	1
21	79	北京交通大学	10004	34	0	1	1

图 5-22 Excel 表中的部分学校数据

然后在软件中右击"表"下面的 school,选择"导入向导"选项,如图 5-23 所示。

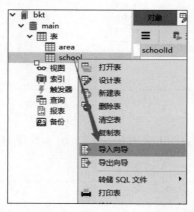

图 5-23 选择"导入向导"选项

进入选择导入格式页面,选择"Excel 文件",然后单击"下一步"按钮,如图 5-24 所示。

进入选择数据源页面,从"导入从"这一栏中选择文件保存地址,这里选择保存在桌面上。然后在"表"框中选择 Sheet1,单击"下一步"按钮,如图 5-25 所示。

进入为源定义一些附加选项的页面,这里一般都采用默认值,直接单击"下一步"按钮便可,如图 5-26 所示。

进入选择目标表页面,这里的目标表就是 SQLite 数据库中的 school 表,因此不用更

第 5 章　数据库开发

图 5-24　选择 Excel 文件格式

图 5-25　选择文件作为数据源

改，单击"下一步"按钮便可，如图 5-27 所示。

进入定义栏位对应页面，"目标栏位"与"源栏位"要一一对应。因为 Excel 表中的列名和数据库 school 表的字段名是相同的，所以好区分。当名称命名不同时一定要注意对应正确。再单击"下一步"按钮，如图 5-28 所示。

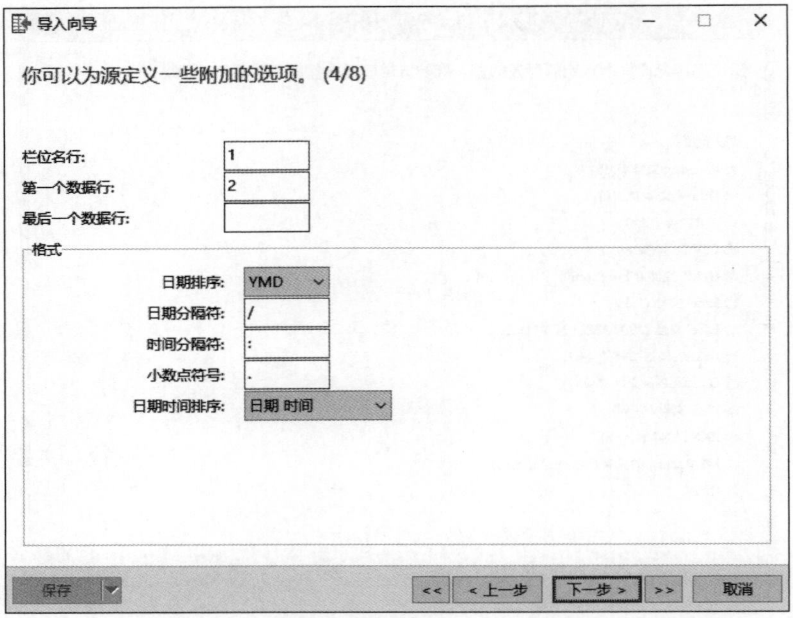

图 5-26　定义附加选项

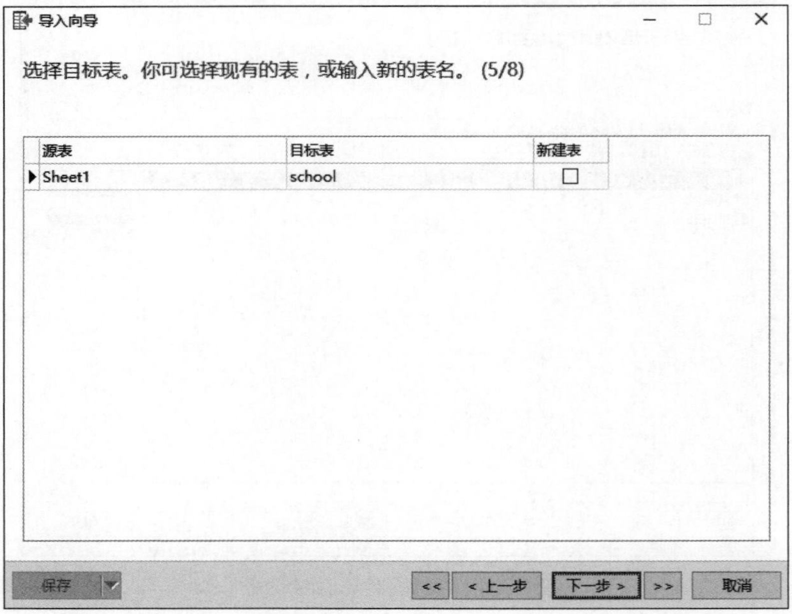

图 5-27　选择目标表

进入选择导入模式页面,这里我们选择"添加:添加记录到目标表"选项,然后单击"下一步"按钮,如图 5-29 所示。

单击"开始"按钮进行导入,完成导入向导,如图 5-30 所示。

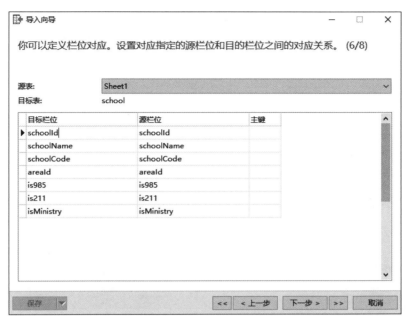

图 5-28　定义栏位对应

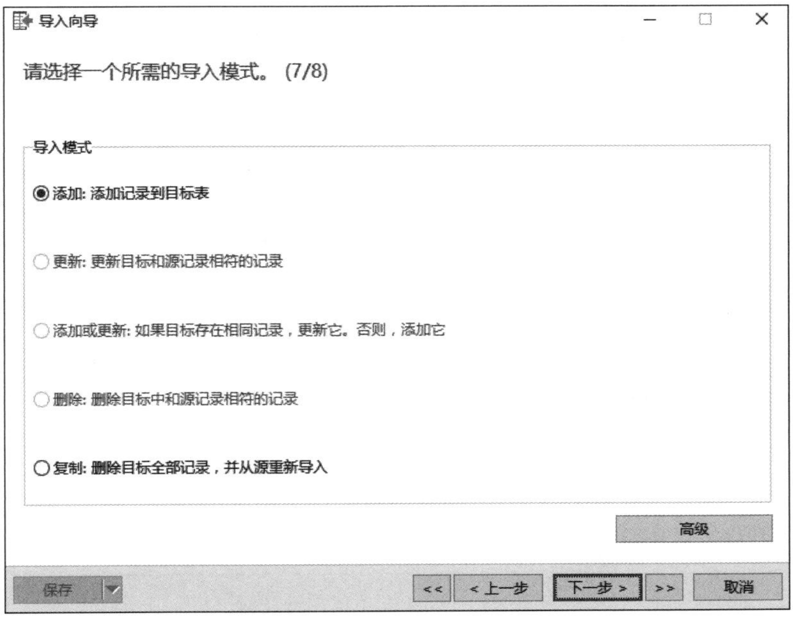

图 5-29　选择导入模式

然后刷新 school 表,可以看到表中已经有了新添加的数据,如图 5-31 所示。

至此,我们已经完整地创建了 SQLite 本地数据库 bkt,它的名称是 bkt.sqlite,如图 5-32 所示。

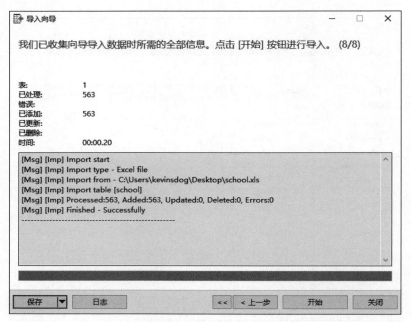

图 5-30　完成导入向导

图 5-31　school 表中的部分记录

图 5-32　bkt.sqlite 本地数据库

在实战中,我们把该文件放到项目目录下便可直接使用(从这个本地数据库读取数据,而不是从数组中取值)。具体放到哪里以及如何使用在 5.5 节会详细介绍。

5.4 SQLite 数据库操作类与接口

刚刚通过一个小案例介绍了如何使用代码创建 SQLite 数据库及表,现在我们将继续详细讲解 SQLite 并实现用代码操作数据库,以及读取本地 SQLite 数据(准备好的 bkt.sqlite)来替换前面章节从数组中获取的数据。首先,我们对 SQLite 要有个详细的了解。

由于在 Android 中集成了 SQLite 数据库,因此每个 Android 应用程序都可以使用 SQLite 数据库。在 SDK 中的 samples/NotePad 下可以找到关于如何使用数据库的例子。

在 Android 系统中,如果要进行 SQLite 数据库的操作,主要使用 Android 系统提供的下面几个类或接口。

- android.database.sqlite.SQLiteDataBase:完成数据的增、删、改、查操作。
- android.database.sqlite.SQLiteOpenHelper:完成对数据库创建和升级。
- android.database.Cursor:保存所有的查询结果。
- android.database.ContentValues:对传递的数据进行封装。

下面我们来学习这几个类。

5.4.1 SQLiteDataBase 类

在 Android 系统中提供了一个名为 SQLiteDataBase 的类,该类封装了一些操作数据库的 API,通过该类可以执行 SQL 语句,以完成对数据表的增加、修改、删除、查询等操作。SQLite 是一个数据库对象,它不仅支持执行 SQL 语句,还提供了一些操作数据库的直接方法。对 SQLiteDataBase 的学习应该重点掌握 execSQL()和 rawQuery()方法。execSQL()方法可以执行 SQL 语句;而 rawQuery()方法用于执行 SELECT 语句。

除了 execSQL()和 rawQuery()方法,SQLiteDataBase 还专门提供了对应于添加、查询、更新、删除的操作方法:inset()、query()、update()、delete()。这些方法主要是给不太了解 SQL 语法的初学者使用的,对于熟悉 SQL 语法的程序员而言,直接使用 execSQL()和 rawQuery()方法执行 SQL 语句便可。实际上,这些方法的内部也是执行 SQL 语句,由系统根据这些参数拼接一个完整的 SQL 语句。

SQLiteDataBase 类定义的常用操作方法如表 5-1 所示。

表 5-1 SQLiteDataBase 类定义的常用操作方法

方　　法	描　　述
execSQL(String sql,Object[] bindArgs)	执行 SQL 语句,同时绑定参数
rawQuery(String sql,String[] selectionArgs)	执行带占位符的 SQL 查询
execSQL(String sql)	执行 SQL 语句

5.4.2 SQLiteOpenHelper 类

为了让我们能够更加方便地管理数据库，Android 专门提供了一个 SQLiteOpenHelper 帮助类，借助这个类可以对数据库进行创建和升级。但它是一个抽象类，因此在我们要使用它时需要创建一个类去继承它，以实现它的一些方法对数据库进行一些操作。

SQLiteOpenHelper 类的常用方法如表 5-2 所示。使用时需要定义其子类，并且在子类中重写相应的抽象方法。

表 5-2　SQLiteOpenHelper 类定义的方法

方　　法	描　　述
SQLiteOpenHelper（Contextcontext，Stringname，SQLiteDataBase.CursorFactory，int version）	通过此构造函数指明要操作的数据库名称以及数据的版本编号
getReadableDatabase()	以只读的方式创建或打开数据库
getWritableDatabase()	以修改的方式创建或打开数据库
onCreate(SQLiteDatabase db)	创建数据表
onOpen(SQLiteDatabase db)	打开数据表
onUpgrade(SQLiteDatabase db，int oldVersion，int new Version)	更新数据表

5.4.3 Cursor 接口

除上述操作之外，数据库还有一项数据的查询操作。当 Android 程序进行数据查询操作时需要保存全部的查询结果，这时就要用到 Cursor。Cursor 也称游标，用于对查询结果进行随机访问，它提供了以下方法来将记录指标移动到相应的记录。

- move(int offset)：将记录指针向上或向下移动指定的行数。offset 为正数就向下移动，为负数就向上移动。
- moveToNext()：将游标从当前记录移动到下一条记录，如果已经移过了结果集的最后一条记录，则返回值为 false，否则为 true。
- moveToPrevious()：将游标从当前记录移动到上一条记录，如果已经移过了结果集的第一条记录，则返回值为 false，否则为 true。
- moveToFirst()：将游标移动到结果集的第一条记录，如果结果集为空，则返回值为 false，否则为 true。
- moveToLast()：将游标移动到结果集的最后一条记录，如果结果集为空，则返回值为 false，否则为 true。

5.4.4 ContentValues 类

前面讲到 SQLiteDataBase 类中提供了增删改查等方法，在使用这些方法进行数据处理时，我们都必须使用 ContentValues 类进行封装。ContentValues 类和 Hashtable 比较类似，它也是负责存储一些键-值对，但是它存储的键-值对中的键是一个 String 类型，而值都是基本类型。ContentValues 类的几个常用方法如下所示。
- ContentValues()：实例化一个 ContentValues 类。
- clear()：清空全部数据。
- put(String key,包装类 value)：设置列名和对应的列值。
- getAs 包装类(String key)：根据 key 值取得数据。

5.5 从 SQLite 获取 ListView 列表项的值

5.5.1 项目结构

图 5-33 所示为 Code0501 项目结构。

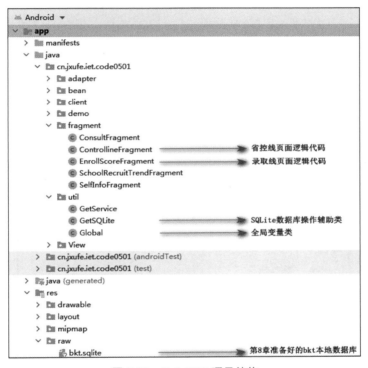

图 5-33　Code0501 项目结构

5.5.2 案例流程模块

通过刚刚的学习，我们已经知道创建 SQLite 数据库以及建表、SQL 操作的具体过程。本节将是这章的重头戏。我们使用代码实现操作 SQLite 数据库的第二种方式，把本节之前实现项目中获取数据的方式进行替换，即把从数组获取数据的方式转换成从本地 SQLite 数据库地区表 area 和学校表 school 中获取。首先，我们先查看哪些地方用到了这两张表的数据。

"历年高考省控线查询""历年录取线查询"两个模块中都使用过，如图 5-34 和图 5-35

所示。

图 5-34 "历年高考省控线查询"页面 　　图 5-35 "历年录取线查询"页面

5.5.3 定义全局变量类

在 util 包下新建一个类 Global,该类用于定义一些全局变量,以方便使用和修改,如图 5-36 所示。

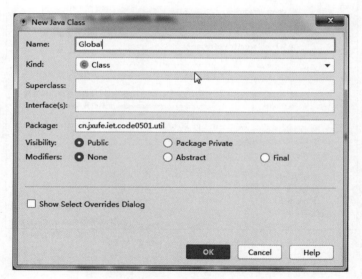

图 5-36 新建 Global 类

Global 类中的代码如程序清单 5-1 所示。

程序清单 5-1：Code0501\app\src\main\java\cn\jxufe\iet\code0501\util\Global.java

```
1    public class Global {                              //定义一些全局变量
2        public static int sourceAreaId=2;              //生源地 Id,默认为江西
3        public static int targetAreaId=2;              //意向省份 Id
```

```
4       public static int schoolAreaId=2;
5       public static int batchId = 1;                      //批次 Id
6       public static int categoryId = 1;                   //科目 Id
7       public static String DB_PATH;                       //数据库的路径
8       public static String DB_NAME = "bkt.db";            //数据库的名称
9       public static String[] batches = new String[] { "不限", "一本", "二本",
    "三本", "专科", "提前" };              //保存所有的批次,数组下标与其对应的 Id 一致
10      public static String[] categories = new String[] { "不限", "文史", "理工",
    "综合类", "艺术类", "体育" };          //保存所有的科类,数组下标与其对应的 Id 一致
11      public static String URL="http://10.0.2.2:8080/bkt/";   //本地服务器地址
12      public static GetSQLite getSQLite;
13      public static SQLiteDatabase db;
14      public static SharedPreferences initPreferences;    //本地保存的信息
15  }
```

5.5.4 数据库操作辅助类

首先,我们在 res 目录下新建一个文件夹,这里取名为 raw 文件夹,其目的是存放本地数据库,如图 5-37 和图 5-38 所示。

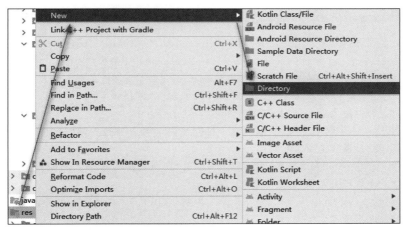

图 5-37 在 res 下新建 Directory

图 5-38 新建 raw 文件夹

然后我们再将准备好的本地 SQLite 文件(在第 5 章使用工具创建好的 bkt.sqlite)复制到项目中,放在 res 目录的 raw 文件夹下,如图 5-39 所示。

图 5-39　布置好本地数据库 bkt.sqlite

　　数据库准备好后,接下来要做的就是编写代码找到这个数据库,并且进行对数据的一系列操作。这里先明确一点,现在是从本地数据库(bkt.sqlite)中取数据,之前我们用到 area 信息和 school 信息时都是从数组中取数据。我们不妨来回忆前面章节是怎么做的。

　　在第 4 章中,生源地数据(area 表)和学校数据(school 表)是临时定义的数据集合,代码如程序清单 5-2 所示。

程序清单 5-2：Code0501\app\src\main\java\cn\jxufe\iet\code0501\util\GetSQLite.java

```
1    public class GetSQLite {
2        private String[] provinces = new String[]{"上海", "浙江", "江西", "北京"};
3        private String[][] schoollists=new String[][]{
4                {"复旦大学","上海中医药大学","同济大学","上海大学","上海交通大学","上海财经大学"},
5                {"浙江大学","杭州电子科技大学","杭州师范大学","浙江财经学院","浙江传媒学院","浙江工商大学"},
6                {"南昌大学","江西师范大学","江西农业大学","江西财经大学","华东交通大学","东华理工大学"},
7                {"北京大学","清华大学","中国人民大学","北京交通大学","北京航空航天大学","北京理工大学"}};
8        /* 把省份数据封装成集合,为后面调用数据库数据做准备 */
9        public List<String> setProvince(){
10           List<String> provinces=new ArrayList<>();
11           provinces.add("上海");
12           provinces.add("浙江");
13           provinces.add("江西");
14           provinces.add("北京");
15           return provinces;
16       }
17       /* 把学校信息封装成集合,为后面调用数据库数据做准备 */
18       public List<School> setSchoolList(){
19           List<School> list=new ArrayList<>();
```

```
20          for (int i=0;i<schoollists.length;i++){
21              for(int j=0;j<schoollists[i].length;j++){
22                  School school=new School();
23                  school.setAreaName(provinces[i]);
24                  school.setSchoolName(schoollists[i][j]);
25                  list.add(school);
26              }
27          }
28      return list;
29      }
30 }
```

这里,我们需要更改 GetSQLite 这个类,让它成为对本地数据库的一个操作辅助类,用于初始化数据库以及打开指定的数据库,也就是我们放在 raw 下面的 bkt.sqlite 文件,并且获取数据库表及其相关信息。其代码如程序清单 5-3 所示。

程序清单 5-3:Code0501\app\src\main\java\cn\jxufe\iet\code0501\util\GetSQLite.java

```
1  public class GetSQLite{
2  public GetSQLite(Context context){
3      Global.initPreferences = context.getSharedPreferences("dbInfo",
4              Context.MODE_PRIVATE);             //初始化数据库保存信息
5      boolean isInitDB=Global.initPreferences.getBoolean("initDB", false);
6      if(!isInitDB){                             //如果没有初始化数据库
7          initDB(context);                       //如果是第一次使用,则执行初始化
8          Global.initPreferences.edit().putString("dbPath", Global.DB_PATH)
.commit();
9      Global.initPreferences.edit().putBoolean("initDB",true).commit();
                                                  //保存数据并提交
10     }else{Global.DB_PATH=Global.initPreferences.getString("dbPath","");
11     }
12     System.out.println("本地数据库路径:"+Global.DB_PATH);
13 }
```

这里我们用构造函数初始化数据库并调用 initDB() 方法,这个方法主要用来判断数据库是否已经初始化,代码如程序清单 5-4 所示。

程序清单 5-4:Code0501\app\src\main\java\cn\jxufe\iet\code0501\util\GetSQLite.java

```
1  public void initDB(Context context) {         //初始化数据库
2      File externalDir = context.getExternalFilesDir(Environment.DIRECTORY_
DOWNLOADS);                                       //存储路径
3      String dbPath = externalDir.getAbsolutePath() + File.separator
           + "db";                                //在该路径下创建一个 db 文件夹
4      File dir = new File(dbPath);
```

```
5       if (!dir.exists()) {                         //如果文件夹不存在,则创建文件夹
6           dir.mkdir();                             //创建文件夹
7       }
8       File file = new File(dir.getPath(), Global.DB_NAME);   //定义数据库文件
9       if (!file.exists()) {                        //如果文件不存在,则创建文件
10          try {
11              file.createNewFile();                //创建文件
12          } catch (IOException e) {
13              System.out.println("创建文件时出错!");
14              e.printStackTrace();
15          }
16          copyFile(context, R.raw.bkt, file);      //复制资源,将 raw 文件夹下的资
                                                     //源复制到存储卡中
17      }
18      Global.DB_PATH=file.getAbsolutePath();       //获取数据库的路径
19  }
```

该方法中还调用了 copyFile() 方法,用于复制资源。将 raw 文件夹下的资源复制到存储卡中,并且调用 Global 类中的 DB_PATH 保存数据库的路径。CopyFile() 方法的代码如程序清单 5-5 所示。

程序清单 5-5:Code0501\app\src\main\java\cn\jxufe\iet\code0501\util\GetSQLite.java

```
1   public void copyFile(Context context, int resId, File file){
                                    //将 R.raw 中的数据库文件复制到存储卡中
2       try {
3           InputStream is =context.getResources().openRawResource(R.raw.bkt);
                                    //打开文件对应的输入流
4           FileOutputStream fos = new FileOutputStream(file);
                                    //打开文件对应的输出流
5           byte[] buffer = new byte[1024];
6           int hasRead = 0;
7           while ((hasRead = is.read(buffer)) != -1) {
8               fos.write(buffer, 0, hasRead);
9           }
10          fos.close();
11          is.close();                              //关闭资源
12      } catch (Exception e) {
13          System.out.println("复制文件时出错啦!");
14      }
15  }
```

初始化数据库并复制完资源后,接下来要对指定的数据库进行一系列的操作。首先打开指定的数据库,代码如程序清单 5-6 所示。

程序清单 5-6：Code0501\app\src\main\java\cn\jxufe\iet\code0501\util\GetSQLite.java

```
1   public SQLiteDatabase getDataBase(Context context) {    //打开指定的数据库
2       ReturnGlobal.db=context.openOrCreateDatabase(Global.DB_PATH,Context
    .MODE_PRIVATE, null); }                                 //打开数据库
```

然后根据需求编写通用方法，可以根据不同的操作编写不同的方法。我们这里有以下几个方法。

获取数据库表中某一字段的值的集合，代码如程序清单 5-7 所示。

程序清单 5-7：Code0501\app\src\main\java\cn\jxufe\iet\code0501\util\GetSQLite.java

```
1   public List<String> getField(Context context, String sql) {
                                        //获取数据库表中某一字段的值的集合
2       List<String> fields = new ArrayList<String>();
3       SQLiteDatabase db=getDataBase(context);
4       Cursor cursor = db.rawQuery(sql, null);
5       while (cursor.moveToNext()) {
6           fields.add(cursor.getString(0));
7       }
8       db.close();
9       return fields;
10  }
```

根据地区获取该地区下的所有学校信息，代码如程序清单 5-8 所示。

程序清单 5-8：Code0501\app\src\main\java\cn\jxufe\iet\code0501\util\GetSQLite.java

```
1   public List<School> getSchoolsByArea(Context context,String sql,String[]
    args){
2       List<School> schools=new ArrayList<School>();
3       SQLiteDatabase db=getDataBase(context);
4       Cursor cursor=db.rawQuery(sql,args);
5       while(cursor.moveToNext()){
6           School school=new School();
7   school.setSchoolName(cursor.getString(cursor.getColumnIndex(School
    .SCHOOL_NAME)));
8       school.setSchoolCode(cursor.getInt(cursor.getColumnIndex(School.SCHOOL_CODE)));
9       school.setSchoolId(cursor.getInt(cursor.getColumnIndex(School.SCHOOL_ID)));
10          schools.add(school);
11      }
12      cursor.close();
13      db.close();
14      return schools;
```

```
15   }
```

根据批次名称获取批次 Id,代码如程序清单 5-9 所示。

程序清单 5-9:Code0501\app\src\main\java\cn\jxufe\iet\code0501\util\GetSQLite.java

```
1   public int getBatchId(String batchName){
2       String[] batchNames=new String[]{"不限", "一本", "二本", "三本", "专科", "提前"};
3       for(int i=0;i<batchNames.length;i++){
4           if(batchName.equals(batchNames[i])){
5               return i;
6           }
7       }
8       return 0;
9   }
```

根据科类名称获取类别 Id,代码如程序清单 5-10 所示。

程序清单 5-10:Code0501\app\src\main\java\cn\jxufe\iet\code0501\util\GetSQLite.java

```
1   public int getCategoryId(String categoryName){
2       String[] categoryNames=new String[]{ "不限", "文史", "理工", "综合类", "艺术类", "体育"};
3       for(int i=0;i<categoryNames.length;i++){
4           if(categoryName.equals(categoryNames[i])){
5               return i;
6           }
7       }
8       return 0;
9   }
```

根据地区名称获取地区 Id,代码如程序清单 5-11 所示。

程序清单 5-11:Code0501\app\src\main\java\cn\jxufe\iet\code0501\util\GetSQLite.java

```
1   public String getAreaName(Context context,String areaId){
2       SQLiteDatabase db=getDataBase(context);
3       Cursor cursor=db.rawQuery("select Name from bkt_area where ID=?",new String[]{areaId});
4       while(cursor.moveToNext()){
5           return cursor.getString(cursor.getColumnIndex("Name"));
6       }
7       db.close();
8       return "";
9   }
```

5.5.5 替换 area 表数据

我们知道在"历年高考省控线查询"页面上有"生源地"控件,如图 5-40 所示。

图 5-40 "省控线查询"页面上的"生源地"控件

之前做的例子都是显示数组中的值,我们也不妨来看前面是怎么做的,代码如程序清单 5-12 所示①。

程序清单 5-12:Code040201\app\src\main\java\cn\jxufe\iet\code040201\GetSQLite.java

```
1   public class GetSQLite {
2       /*把省份数据封装成集合,为后面调用数据库数据做准备*/
3       public List<String> setProvince(){
4           List<String> provinces=new ArrayList<>();
5           provinces.add("上海");
6           provinces.add("浙江");
7           provinces.add("江西");
8           provinces.add("北京");
9           return provinces;
10      }
11  }
```

在 ControlLineActivity 主类文件的 onCreate 方法中,实例化 GetSQLite 对象获取本地数据临时生源地集合。然后,定义 4 个 Spinner 列表适配器并将数据源和布局关联,这样就实现了 4 个 Spinner 列表,代码如程序清单 5-13 所示。

程序清单 5-13:Code040201\app\src\main\java\cn\jxufe\iet\code040201\ControlLineActivity.java

```
1   GetSQLite getSQLite=new GetSQLite();
2   provinces=getSQLite.setProvince();                    //获取省份集合
3   ArrayAdapter<String> provincesAdapter = new ArrayAdapter<String>(
4       this, R.layout.spinner_item_year, provinces);        ——从数据中取值
5   provinceSpinner.setAdapter(provincesAdapter);
6   provinceSpinner.setSelection(sourceAreaId);           //设置默认省份显示项
7   ArrayAdapter<String> yearAdapter = new ArrayAdapter<String>(
8       this, R.layout.spinner_item_year, years);
9   yearSpinner.setAdapter(yearAdapter);
10  ArrayAdapter<String> categoryAdapter = new ArrayAdapter<String>(
```

---

① 本小节以后集成了第 4 章的部分代码,这部分代码以 Code04 开头,请读者注意。

```
11          this, R.layout.spinner_item_category,categorys);
12  categorySpinner.setAdapter(categoryAdapter);
13  categorySpinner.setSelection(categoryId);              //设置默认科类显示项
14  ArrayAdapter<String> batchAdapter = new ArrayAdapter<String>(
15          this, R.layout.spinner_item_batch, batchs);
16  batchSpinner.setAdapter(batchAdapter);
17  batchSpinner.setSelection(batchId);                    //设置默认批次显示项
```

现在我们改成从本地数据库获取，打开 ControllineFragment 类（见程序清单 5-14）。

程序清单 5-14：Code040201\app\src\main\java\cn\jxufe\iet\code040201\fragment\ControllineFragment.java

```
1  GetSQLite getSQLite = new GetSQLite();         ——→从数组中取值
2  provinces = getSQLite.setProvince();           //获取省份集合
```

这里是调用 GetSQLite 类的数组中的值，将它改成如程序清单 5-15 所示。

程序清单 5-15：Code0501\app\src\main\java\cn\jxufe\iet\code0501\fragment\ControllineFragment.java

```
1  GetSQLite getSQLite=new GetSQLite(getActivity());
2      provinces = getSQLite.getField(getActivity(), "select areaName from
   area");                    //获取所有的省份         ——→从本地数据库取值
3  ArrayAdapter<String> provincesAdapter = new ArrayAdapter<String>(
4          getActivity(), R.layout.spinner_item_year, provinces);
5  provinceSpinner.setAdapter(provincesAdapter);
```

这里创建一个 GetSQLite 对象并调用其中的 getField() 方法，读取 area 表中 areaName 的值并加载到 provincesAdapter 适配器中，最后显示在列表上。看上去是不是很简单？注意，修改代码时只要找到与 area 有关的代码替换便可，其他代码不能动。

在"历年录取线查询"页面也有与 area 有关的控件，分别是"学校省份"和"生源地"，如图 5-41 和图 5-42 所示。

图 5-41 学校省份控件

图 5-42 生源地控件

首先，我们来复习第 4 章的 Code040203 中是如何定义 GetSQLite 类且从数组中获取数据的，其代码如程序清单 5-16 所示。

程序清单 5-16：Code040203\app\src\main\java\cn\jxufe\iet\code040203\EnrollScoreActivity.java

```
1  GetSQLite getSQLite=new GetSQLite();           //创建一个 GetSQLite 对象
2  provinces=getSQLite.setProvince();             //获取省份集合
3  ArrayAdapter<String> areaAdapter = new ArrayAdapter<String>
```

```
4        (EnrollScoreActivity.this, R.layout.spinner_item_year, provinces);
5    sourceAreaSpinner.setAdapter(areaAdapter);     //学校所在地列表
6    schoolAreaSpinner.setAdapter(areaAdapter);     //生源地所在地列表
7    ArrayAdapter<String> categoryAdapter = new ArrayAdapter<String>(
8        EnrollScoreActivity.this, R.layout.spinner_item_category, categories);
9    categorySpinner.setAdapter(categoryAdapter);   //科类列表
10   ArrayAdapter<String> batchAdapter = new ArrayAdapter<String>(
11       EnrollScoreActivity.this, R.layout.spinner_item_batch, batches);
12   batchSpinner.setAdapter(batchAdapter);         //批次列表
13   batchSpinner.setSelection(batchId - 1);        //默认显示项
14   ArrayAdapter<String> yearAdapter = new ArrayAdapter<String>(
15       EnrollScoreActivity.this, R.layout.spinner_item_year, years);
16   yearSpinner.setAdapter(yearAdapter);           //年份列表
17   yearSpinner.setSelection(years.length - 1);    //默认显示项
```

其中,从数组中取值的代码如程序清单 5-17 所示。

程序清单 5-17:Code040203\app\src\main\java\cn\jxufe\iet\code040203\EnrollScoreActivity.java

```
GetSQLite getSQLite = new GetSQLite();         //从数组中取值
provinces = getSQLite.setProvince();           //获取省份集合
```

这里是调用 GetSQLite 类的数组中的值,现在改成从本地数据库获取,打开 EnrollScoreFragment 类,代码如程序清单 5-18 所示。

程序清单 5-18:Code0501\app\src\main\java\cn\jxufe\iet\code0501\fragment\EnrollScoreFragment.java

```
1   Private List<String>provinces = new ArrayList<String>();
2   public void onCreate(Bundle savedInstanceState) {
3       super.onCreate(savedInstanceState);   //从本地数据库取值
4       GetSQLite getSQLite=new GetSQLite(getActivity());
5       provinces = getSQLite.getField(getActivity(),
6           "select areaName from area");      //获取所有的省份
7       System.out.println("provinces="+provinces);
8   }
```

这里我们先定义省份的集合 provinces,然后将本地数据库 area 表的数据都保存在 provinces 中,如果哪个列表控件要显示 area 表的信息,只要把 provinces 加载到列表的适配器上。因此下面依次修改学校省份和生源地的相关代码,如程序清单 5-19 所示。

程序清单 5-19:Code0501\app\src\main\java\cn\jxufe\iet\code0501\fragment\EnrollScoreFragment.java

```
1   ArrayAdapter<String> areaAdapter = new ArrayAdapter<String>(getActivity(),
        R.layout.spinner_item_year, provinces);
2       sourceAreaSpinner.setAdapter(areaAdapter);     //学校所在地列表
3       schoolAreaSpinner.setAdapter(areaAdapter);     //生源地所在地列表
```

定义 areaAdapter 适配器,将集合 provinces 加载到适配器中,然后再将适配器赋给两个控件列表。

5.5.6 替换 school 表数据

在"历年录取线查询"页面上有"选择学校"控件,这里是我们需要替换的地方,如图 5-43 所示。

先要明确一点,这里显示的信息并不是所有学校,而是位于对应省份的学校,因此不能像替换 area 表数据一样直接查询出来然后赋值。我们应该写一个方法,找出某省份下的所有学校信息再去赋值。

图 5-43 "选择学校"控件

我们前面更改 GetSQLite 类时新增加了几个对数据库操作的方法,其中有一个方法就是根据地区获取该地区下的所有学校信息,代码如程序清单 5-20 所示。

程序清单 5-20:Code0501\app\src\main\java\cn\jxufe\iet\code0501\util\GetSQLite.java

```
1   public List< School > getSchoolsByArea(Context context, String sql, String[] args){
2       List<School> schools=new ArrayList<School>();
3       SQLiteDatabase db=getDataBase(context);
4       Cursor cursor=db.rawQuery(sql,args);
5       while(cursor.moveToNext()){
6           School school=new School();
7   school.setSchoolName(cursor.getString(cursor.getColumnIndex(School.SCHOOL_NAME)));
8       school.setSchoolCode(cursor.getInt(cursor.getColumnIndex(School.SCHOOL_CODE)));
9       school.setSchoolId(cursor.getInt(cursor.getColumnIndex(School.SCHOOL_ID)));
10          schools.add(school);
11      }
12      cursor.close();
13      db.close();
14      return schools;
15  }
```

在之前章节中,我们从集合里面获取值时已经写了一个 getSchool 方法,其代码如程序清单 5-21 所示。

程序清单 5-21:Code040203\app\src\main\java\cn\jxufe\iet\code040203\EnrollScoreActivity.java

```
1   public   void getSchool(String areaName){
2       GetSQLite getSQLite=new GetSQLite();
3       schoolList=getSQLite.setSchoolList();
```

```
4        schools.clear();                                      //学校名称集合清空
5        for (School school : schoolList) {
6            if(school.getAreaName().equals(areaName)){   //寻找对应省份的学校
7                schools.add(school.getSchoolName());//把数据添加进学校名称集合
8            }
9        }
10       ArrayAdapter<String> schoolAdapter = new ArrayAdapter<String>(
11               EnrollScoreActivity.this, R.layout.spinner_item_year, schools);
12       schoolSpinner.setAdapter(schoolAdapter);
13   }
```

现在我们需要重新编写这个方法，打开 EnrollScoreFragment 类，改写 getSchool() 方法，代码如程序清单 5-22 所示。

程序清单 5-22：Code0501\app\src\main\java\cn\jxufe\iet\code0501\fragment\EnrollScoreFragment.java

```
1    public  void getSchool(String areaName){
2        //GetSQLite getSQLite =new GetSQLite();
3    GetSQLite getSQLite=new GetSQLite(getActivity());
4        schoolList=getSQLite.getSchoolsByArea(getActivity(),"select * from
                school s,area a where s.areaId=a.areaId and a.areaName=?",
5               new String[] { areaName });
6        schools.clear();                                      //学校名称集合清空
7        for (School school : schoolList) {
8    //       if(school.getAreaName().equals(areaName)){//寻找对应省份的学校
9                schools.add(school.getSchoolName());         //把数据添加进学校名称集合
10   //       }
11       }
12       ArrayAdapter<String> schoolAdapter = new ArrayAdapter<String>(
13               getActivity(), R.layout.spinner_item_year, schools);
14       schoolSpinner.setAdapter(schoolAdapter);
15   }
```

这里只需要修改 getSchool() 方法便可。然后运行程序，可以看到替换数据成功，如图 5-44 和图 5-45 所示。

图 5-44 替换 area 表数据的效果

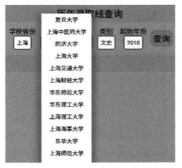

图 5-45 替换 school 表数据的效果

5.6 MySQL 数据库的构建

在"豹考通"应用案例中,选择的数据库系统是 MySQL。MySQL 是一个开源的小型关系数据库管理系统,开发者为瑞典的 MySQL AB 公司。MySQL 被广泛应用在 Internet 上的中小型网站中。由于其体积小、速度快、总体拥有成本低,尤其是开源这一特点,许多中小型网站为了降低网站总体拥有成本而选择 MySQL 作为网站的数据库。

"豹考通"项目主要的数据存储和处理操作放在服务器端,Android 客户端主要用于发送请求和显示操作结果。因此,核心的数据库放在服务器端,采用 MySQL 数据库存储;Android 客户端存储少量的相对固定且操作较为频繁的数据,例如省份信息、省份下的学校信息等,直接使用内嵌的小型关系数据库 SQLite 保存。为了使本地数据库与服务器端数据内容保持一致,本地数据库中的表的结构及其内容直接从服务器端导出。

在建好表以后,可以用 SQL 语句进行一些简单的插入数据等操作,但一张表中存的数据肯定不止这么多,那么成百上千条记录的数据该如何放进表中呢?本节将介绍几种导入数据的方法。

5.6.1 MySQL 的应用范围

主要适用场景包括:Web 网站系统(Web 站点是 MySQL 最大的客户群,也是 MySQL 发展史的中流砥柱,因为它安装配置简单和开源免费)、数据仓库系统、日志记录系统、嵌入式系统(嵌入式环境对软件系统最大的限制是硬件资源非常有限,而 MySQL 在硬件资源的使用方面可伸缩性非常强,而且它有专门针对嵌入式环境的版本)等。

小型互联网企业一般都用 MySQL 存储数据,主要原因是免费、架构科学、性能够用。但对于企业中的一些管理软件,有的不会用 MySQL,而用企业级数据库,例如 Oracle、DB2、SQL Server、Sybase 等。

5.6.2 MySQL 的优缺点

MySQL 的优点包括:
- 使用的核心线程是完全多线程的,支持多处理器。
- 有多种类型:1、2、3、4 和 8 字节长度有符号/无符号整数以及 FLOAT、DOUBLE、CHAR、VARCHAR、TEXT、BLOB、DATE、TIME、DATETIME、TIMESTAMP、YEAR 和 ENUM 类型。
- 通过一个高度优化的类库来实现 SQL 函数并像它们能达到的一样快速。通常在查询初始化后不该有任何内存分配,没有内存漏洞。
- 全面支持 SQL 的 GROUP BY 和 ORDER BY 子句,支持聚合函数(COUNT()、COUNT(DISTINCT)、AVG()、STD()、SUM()、MAX() 和 MIN())。可以在同一查询中混合来自不同数据库的表。
- 支持 ANSI SQL 的 LEFT OUTER JOIN 和 ODBC。
- 所有列都有默认值。可以用 INSERT 插入一个表列的子集,那些没用明确给定值的列设置为它们的默认值。

MySQL 的缺点包括：

- 安全系统复杂而非标准，另外只有在调用 MySQL Admin 重读用户权限时才发生改变。
- 缺乏标准的 RI（Referential Integrity）机制。RI 机制（在给定字段域上的一种固定的范围限制）的缺乏可以通过大量的数据类型来补偿。
- 没有一种存储过程语言，这是对习惯于企业级数据库的程序员的最大限制。
- 不支持热备份。
- 价格随平台和安装方式变化。Linux 的 MySQL 如果由用户自己或系统管理员而不是第三方安装，则是免费的，否则必须付许可费。UNIX 或 Linux 自行安装免费，第三方安装需要 200 美元。

MySQL 可以工作在不同的平台上，支持 C、C++、Java、Perl、PHP、Python 和 TCL API。

5.6.3 MySQL 安装

首先找到下载好的 MySQL 安装文件，解压后双击 Setup.exe 文件，弹出安装向导对话框，如图 5-46 所示。

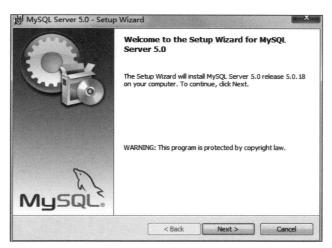

图 5-46　MySQL 安装向导

选择安装类型，这里选择自定义安装，如图 5-47 所示。

下一步弹出选择安装目录的对话框，路径可任意指定，在此将其安装到 D 盘，如图 5-48 所示。

下一步弹出登录对话框，没有账号可直接跳过，如图 5-49 所示。

下一步即完成 MySQL 安装，如图 5-50 所示。

单击 Finish 按钮，将会启动 MySQL 服务器的配置向导，如图 5-51 所示。

单击 Next 按钮，弹出配置和模式选择对话框，配置过程如图 5-52～图 5-62 所示。

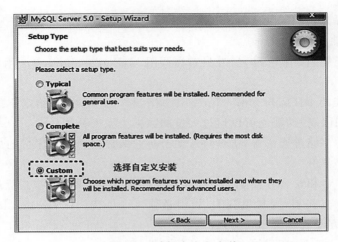

图 5-47　选择自定义安装

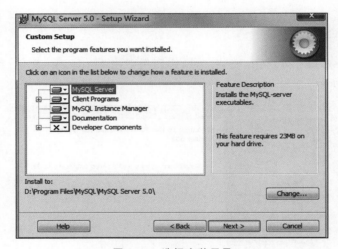

图 5-48　选择安装目录

图 5-49　跳过登录

第 5 章 数据库开发

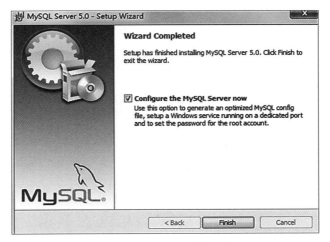

图 5-50　MySQL 安装结束

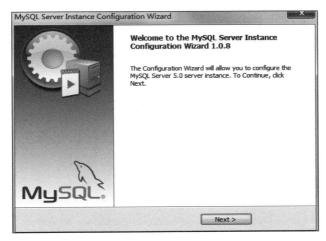

图 5-51　MySQL 配置向导

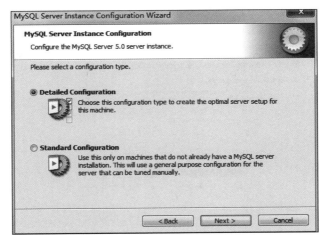

图 5-52　选择详细配置

175

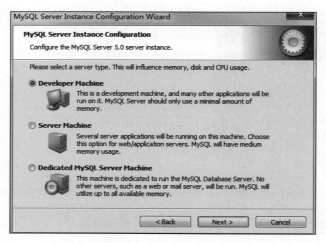

图 5-53　选择开发者模式

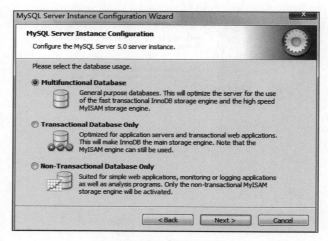

图 5-54　选择多功能数据库

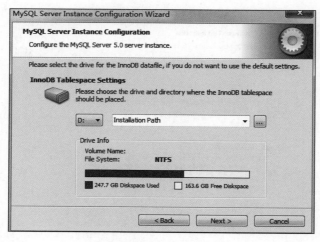

图 5-55　选择驱动存放目录

图 5-56　勾选"命令行"

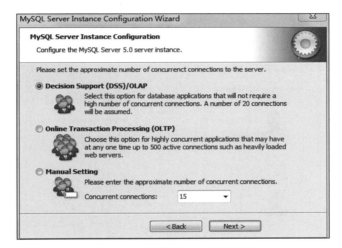

图 5-57　设置并发连接的数目

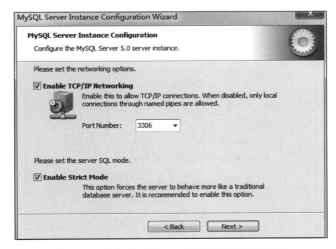

图 5-58　设置 MySQL 的端口号（采用默认值）

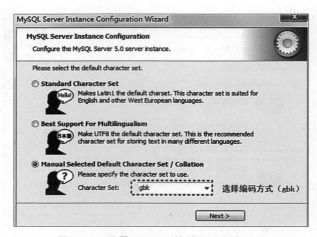

图 5-59　设置 MySQL 的编码（采用 gbk）

注意：默认的是 latin1 不支持中文。

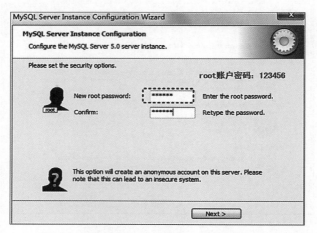

图 5-60　设置 root 账户密码（此处为 123456）

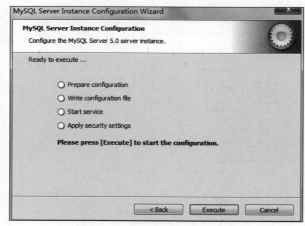

图 5-61　执行刚才设置的配置

第 5 章　数据库开发

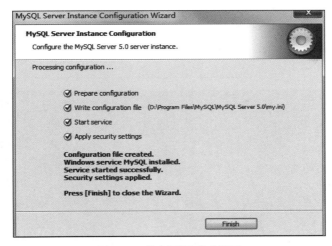

图 5-62　执行配置成功界面

MySQL 安装并配置完成后，单击命令行，输入密码，出现图 5-63 所示的信息，表明 MySQL 安装成功，能够正常使用。

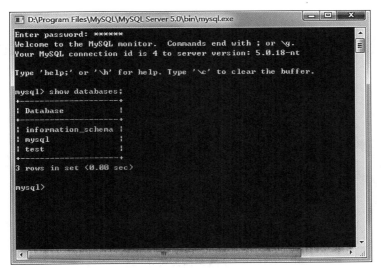

图 5-63　MySQL 命令行信息

5.7　数据库可视化工具安装

安装完 MySQL 后，如何查看表的结构和具体数据呢？如果每次都从命令提示符界面进入 MySQL 进行一些查询、修改等操作，则不仅效率低，而且体验感也非常差。为此，我们推荐一款数据库可视化工具 Navicat for MySQL，本节详细讲解 Navicat 的安装与使用。

5.7.1　Navicat for MySQL 下载

进入官网 https://www.navicat.com.cn/，如图 5-64 所示。

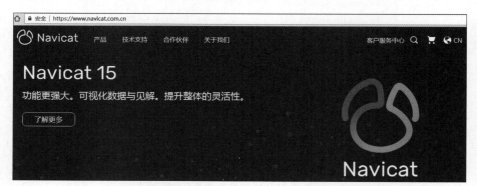

图 5-64　Navicat for MySQL 官网

单击产品选项卡，进入下载页面 https://www.navicat.com.cn/products，如图 5-65 所示。

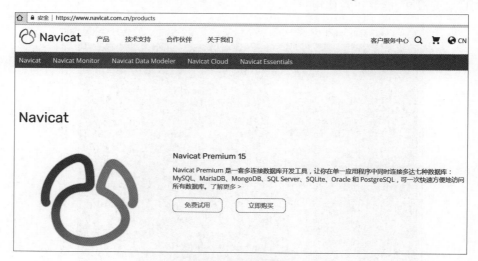

图 5-65　进入下载页面

单击"免费试用"按钮，根据计算机系统位数选择版本，如图 5-66 和图 5-67 所示。

图 5-66　选择免费试用

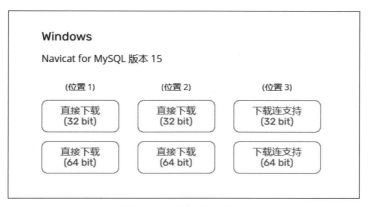

图 5-67　选择版本信息

5.7.2　Navicat for MySQL 安装

下载完成后保存在系统盘中,这里保存在 D 盘,如图 5-68 所示。

图 5-68　保存在 D 盘

双击安装文件,然后单击"下一步"按钮,如图 5-69 所示。

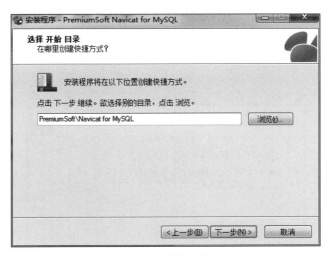

图 5-69　开始安装

选择"我同意"选项,然后单击"下一步"按钮,如图 5-70 所示。
选择安装文件夹,然后单击"下一步"按钮,如图 5-71 所示。

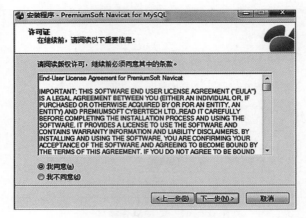

图 5-70 选择"我同意"选项

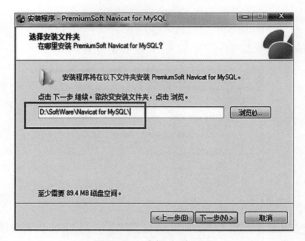

图 5-71 选择安装目录

选择开始目录,如图 5-72 所示。

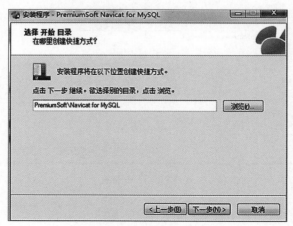

图 5-72 选择开始目录

选择是否创建桌面图标,建议保留默认值,单击"下一步"按钮,如图 5-73 所示。

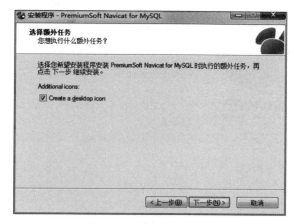

图 5-73　选择额外任务

单击"安装"按钮,如图 5-74~图 5-76 所示。

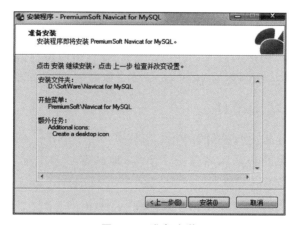

图 5-74　准备安装

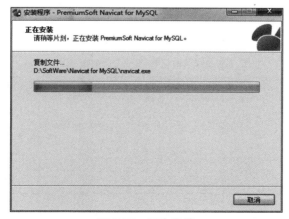

图 5-75　正在安装

图 5-76 完成安装

5.8 数据库表设计与数据的导入

根据第 2 章所述的 E-R 图,可归纳出"豹考通"项目中应包含以下数据库表:地区表、批次表、类别表、学校表、专业表、省控线表、学校录取线表、专业录取线表、问题表。

5.8.1 各表的结构设计

各表的具体字段及其类型说明如下。

地区表包含地区编号和地区名称两个字段,地区编号为主键,其他信息如表 5-3 所示。表 5-4 为本例中地区表的数据记录。

表 5-3 地区表(area)结构

编号	名称	字段代码	类型	长度	是否可空	主键	备注
1	地区编号	areaId	smallint	3	非空	是	
2	地区名称	areaName	varchar	10	非空	否	

表 5-4 地区表(area)数据记录

areaId	areaName
3	上海
10	浙江
13	江西
34	北京

批次表包含批次编号和批次名称两个字段,批次编号为主键,批次名称主要有一本、

二本、三本、专科和提前,其他信息如表 5-5 所示。表 5-6 为本例中批次表的数据记录。

表 5-5　批次表(batch)结构

编号	名称	字段代码	类型	长度	是否可空	主键	备注
1	批次编号	batchId	smallint	3	非空	是	
2	批次名称	batchName	varchar	10	非空	否	

表 5-6　批次表(batch)数据记录

batchId	batchName	batchId	batchName
1	一本	4	专科
2	二本	5	提前
3	三本		

类别表包含类别编号和类别名称两个字段,类别编号为主键,类别名称主要有文史、理工、综合类、艺术类、体育,其他信息如表 5-7 所示。表 5-8 为本例中类别表的数据记录。

表 5-7　类别表(category)结构

编号	名称	字段代码	类型	长度	是否可空	主键	备注
1	类别编号	categoryId	smallint	3	非空	是	自增
2	类别名称	categoryName	varchar	10	非空	否	

表 5-8　类别表(category)数据记录

categoryId	categoryName	categoryId	categoryName
1	文史	4	艺术类
2	理工	5	体育类
3	综合		

学校表包含学校编号、学校名称、学校代码、学校所在地、是否 211、是否 985、是否直属等字段,其中学校编号为主键,学校所在地为外键,其他信息如表 5-9 所示。表 5-10 为本例中学校表的部分数据记录。

表 5-9　学校表(school)结构

编号	名称	字段代码	类型	长度	是否可空	主键	备注
1	学校编号	schoolId	smallint	5	非空	是	自增
2	学校名称	schoolName	varchar	10	非空	否	
3	学校代码	schoolCode	varchar	6	允许空	否	

续表

编号	名称	字段代码	类型	长度	是否可空	主键	备注
4	学校所在地	areaId	smallint	3	非空	否	外键
5	是否211	is211	boolean		非空	否	默认 false
6	是否985	is985	boolean		非空	否	默认 false
7	是否直属	isMinistry	boolean		非空	否	默认 false

表 5-10 学校表（school）部分数据记录

schoolId	schoolName	schoolCode	areaId	is985	is211	isMinistry
5	北京大学	10001	34	1	1	1
6	清华大学	10003	34	1	1	1
8	浙江大学	10335	10	1	1	1
14	复旦大学	10246	3	1	1	1
22	同济大学	10247	3	1	1	1
27	中国人民大学	10002	34	1	1	1
28	北京理工大学	10007	34	1	1	0
31	上海大学	10280	3	0	1	1
39	中央财经大学	10034	34	1	1	1
44	上海交通大学	10248	3	1	1	1
45	上海财经大学	10272	3	0	1	1
48	南昌大学	10403	13	0	1	0
51	北京师范大学	10027	34	1	1	1
59	华东师范大学	10269	3	1	1	1
61	北京邮电大学	10013	34	0	1	1
68	华东理工大学	10251	3	0	1	1
75	中国传媒大学	10033	34	0	1	1
79	北京交通大学	10004	34	0	1	1
86	北京科技大学	10008	34	0	1	1
95	上海理工大学	10252	3	0	0	0
97	江西财经大学	10421	13	0	0	0
99	北京工业大学	10005	34	0	0	0

专业表包含专业编号、专业名称、专业代码等字段，其中专业编号为主键，其他信息如表 5-11 所示。表 5-12 为本例中专业表的部分数据记录。

表 5-11 专业表(major)结构

编号	名称	字段代码	类型	长度	是否可空	主键	备注
1	专业编号	majorId	int	10	非空	是	
2	专业名称	majorName	varchar	20	非空	否	
3	专业代码	majorCode	varchar	6	允许空	否	

表 5-12 专业表(major)部分数据记录

majorId	majorName	majorCode
1	历史学	060101
2	工商管理	120201
3	会计学	120203
4	中国语言文学	
5	法学	030101
6	政治学	
7	财政学	020201
8	国际经济与贸易	020401
9	哲学	010101
10	英语	050201
11	新闻传播学	
12	日语	050207
13	汉语言文学	050101
14	社会学	030301
15	公共管理	
16	统计学	071201
17	经济学	020101
18	国际政治	030202
19	人类学	030303
20	政治学与行政学	030201
21	金融学	020301
22	法语	050204
23	财务管理	120204
24	旅游管理	120901
25	生物科学	071001

续表

majorId	majorName	majorCode
26	管理科学	120101
27	临床医学	100201
28	预防医学	100401
29	海洋科学	070701
30	软件工程	080902
31	土建	
32	化工与制药	
33	航空航天	
34	物理学	070201
35	材料	

省控线表包含省控线编号、省控线值、年份、批次、类别、生源地等字段,其中省控线编号为主键,批次、科类、生源地为外键,其他信息如表5-13所示。表5-14为本例中省控线表的部分数据记录。

表5-13 省控线表(control_line)结构

编号	名称	字段代码	类型	长度	是否可空	主键	备注
1	省控线编号	controlId	int	10	非空	是	自增
2	省控线值	controlLine	smallint	5	允许空	否	
3	年份	controlYear	varchar	4	非空	否	
4	批次	batchId	smallint	3	非空	否	外键
5	科类	categoryId	smallint	3	非空	否	外键
6	生源地	areaId	smallint	3	非空	否	外键

表5-14 省控线表(control_line)部分数据记录

controlId	controlLine	controlYear	categoryId	batchId	areaId
224	400	2010	2	2	3
225	404	2010	1	2	3
243	462	2011	2	1	3
245	468	2011	1	1	3
246	342	2012	2	2	3
247	379	2012	1	2	3
250	151	2012	2	4	3

续表

controlId	controlLine	controlYear	categoryId	batchId	areaId
259	194	2013	1	4	3
260	405	2013	2	1	3
261	448	2013	1	1	3
879	402	2010	2	2	10
880	459	2010	1	2	10
910	593	2012	2	1	10
912	606	2012	1	1	10
913	438	2013	2	2	10
914	468	2013	1	2	10

学校录取线表包含录取线编号、学校、年份、批次、类别、生源地、录取平均分、录取最高分、投档线、最低排名等字段,其中录取线编号为主键,学校、批次、科类、生源地为外键,其他信息如表5-15所示。表5-16为本例中学校录取线表的部分数据记录。

表 5-15 学校录取线表（school_recruit）结构

编号	名称	字段代码	类型	长度	是否可空	主键	备注
1	录取线编号	sRecruitId	int	10	非空	是	自增
2	学校	schoolId	smallint	5	非空	否	外键
3	年份	sRecruitYear	smallint	5	非空	否	
4	批次	batchId	smallint	5	非空	否	外键
5	科类	categoryId	smallint	5	非空	否	外键
6	生源地	areaId	smallint	5	非空	否	外键
7	录取平均分	sRecruitAScore	smallint	5	允许空	否	
8	录取最高分	sRecruitHscore	smallint	5	允许空	否	
9	投档线	enterLine	smallint	5	允许空	否	
10	最低排名	enterLineRank	smallint	5	允许空	否	

表 5-16 学校录取线表（school_recruit）部分数据记录

sRecruitId	schoolId	sRecruitYear	batchId	categoryId	areaId	sRecruitAScore	sRecruitHScore	enterLine
2392	5	2013	1	1	34	604	659	0
2393	5	2012	1	1	34	622	670	0
2394	5	2011	1	1	34	645	675	0

专业录取线表包含录取线编号、学校、专业、年份、批次、科类、生源地、录取平均分、录取最高分等字段,其中录取线编号为主键,学校、专业、批次、科类、生源地为外键,其他信息如表5-17所示。表5-18为本例中专业录取线表的部分数据记录。

表5-17 专业录取线表(major_recruit)结构

编号	名 称	字段代码	类 型	长度	是否可空	主键	备 注
1	录取线编号	majorRecruitId	int	10	非空	是	自增
2	学校	schoolId	smallint	5	非空	否	外键
3	专业	majorId	int	10	非空	否	外键
4	年份	year	varchar	4	非空	否	
5	批次	batchId	smallint	5	非空	否	外键
6	科类	categoryId	smallint	5	非空	否	外键
7	生源地	areaId	smallint	5	非空	否	外键
8	录取平均分	mRecruitAScore	smallint	5	允许空	否	
9	录取最高分	mRecruitHScore	smallint	5	允许空	否	

表5-18 专业录取线表(major_recruit)部分数据记录

majorRecruitId	majorId	schoolId	year	categoryId	batchId	areaId	mRecruitHScore	mRecruitAScore
212	57	8	2013	1	1	34	646	635
226	50	8	2013	1	1	34	593	593
685	10	6	2010	1	1	34	627	627
691	103	6	2010	1	1	34	659	656
705	2	6	2010	1	1	34	650	648
710	58	6	2010	1	1	34	641	639
723	50	6	2010	1	1	34	645	638
736	57	6	2010	1	1	34	649	638
761	5	6	2010	1	1	34	647	635
780	57	6	2011	1	1	34	634	633
792	2	6	2011	1	1	34	650	647

问题表包含问题编号、问题内容、提问时间、回复内容、回复时间、问题状态等字段,其中问题编号为主键,其他信息如表5-19所示。表5-20为本例中问题表的部分数据记录。

表 5-19 问题表(question)

编号	名 称	字段代码	类 型	长度	是否可空	主键	备 注
1	问题编号	questionId	int	10	非空	是	
2	问题内容	questionContent	varchar	100	非空	否	
3	提问时间	questionTime	varchar	20	允许空	否	
4	回复内容	answerContent	varchar	100	允许空	否	
5	回复时间	answerTime	varchar	100	允许空	否	
6	问题状态	questionStatus	varchar	20	允许空	否	

表 5-20 问题表(question)数据记录

questionId	questionContent	questionTime	answerContent	answerTime	questionStatus
1	test	2016-01-25 02:49	1234678	2016-01-25 04:02	未审核
2	hello	2016-01-25 02:53	haha♯www♯ttt	2016-01-25 03:38	未审核
3	get	2016-01-25			暂未回复

5.8.2 建库和建表操作

前面我们已经设计好各个表的结构,包含表的名称、字段代码、类型、长度、主外键等。现在要把这些设计好的结构建立起来。打开 Navicat for MySQL 软件,选中"连接"按钮(见图 5-77)。进入"新建连接"界面,"连接名"这栏可以随意命名,这里输入 bkt。"密码"栏是安装 MySQL 时自己设定的密码,这里为 123456,如图 5-78 所示。

图 5-77 选中"连接"按钮

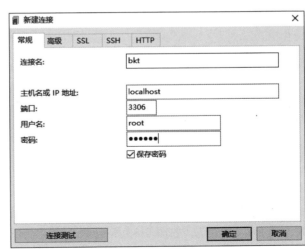

图 5-78 新建数据库连接界面

输入完连接名和密码之后,单击"确定"按钮,可以发现"连接"图标下方出现了名为 bkt 的新连接,如图 5-79 所示。

然后双击 bkt 新连接,右键选择"新建数据库"选项,如图 5-80 所示。

图 5-79　bkt 数据连接

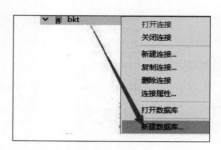

图 5-80　新建数据库

这里数据库名也写为 bkt,字符集选择"utf8--UTF-8 Unicode",排序规则选择"utf8_general_ci",如图 5-81 所示。

然后单击"确定"按钮即可。到这一步,我们已经新建了一个名为 bkt 的数据库连接,又新建了一个名为 bkt 的数据库。接下来,我们要在新建的数据库中新建需要的数据库表。这里以科类表(category)为例讲解如何新建数据表和数据字段以及设置数据类型、长度和大小。其他几张表的操作步骤都基本一致。

首先,选择 bkt 数据库下的"表"选项,右击选择"新建表"选项,如图 5-82 所示。

图 5-81　新建 bkt 数据库　　　　图 5-82　新建表

右边出现建表的界面,然后再根据数据库表的设计在新建的表中添加"名""类型""长度"和"不是 null"等信息,如图 5-83 所示。

科类表主要有两个属性:科类编号和科类名称。这里定义的科类编号不能为空并且自动递增,设置为主键,类型为 smallint。科类名称的类型为 varchar,长度为 10。然后保存信息,可以同时按下 Ctrl+S 键,弹出如图 5-84 所示的显示框。

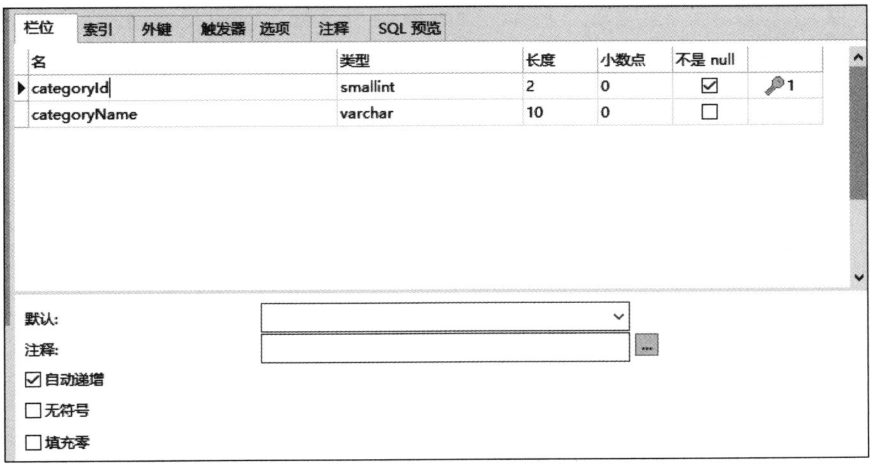

图 5-83　新建科类表(category)

图 5-84　输入表名的对话框

输入表名 category 后单击"确定"按钮即可,其他表的建表过程大体相同,这里不准备一一列出。

5.8.3　使用 SQL 语句建表

除了通过操作步骤建表之外,还可以使用 SQL 语句建表。我们先简单了解什么是 SQL 语句。

SQL(Structured Query Language,结构化查询语言)是关系数据库的标准语言,是一个通用的、功能极强的关系数据库语言。其功能并不只是查询。当前,几乎所有的关系数据库管理系统软件都支持 SQL,许多软件厂商对 SQL 基本命令集还进行了不同程度的扩充和修改。

接下来了解如何使用 SQL 语句建表,同样接着上面讲的操作步骤进行。单击上方的"查询"按钮,出现如图 5-85 所示的界面后单击"新建查询"选项。

然后编写 SQL 语句新建一张科类表,如图 5-86 所示。

单击"运行"按钮后便生成一张新表。这里,我们先了解定义表的语句表示什么意思。SQL 语言使用

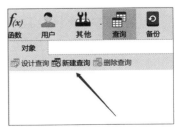

图 5-85　新建查询界面

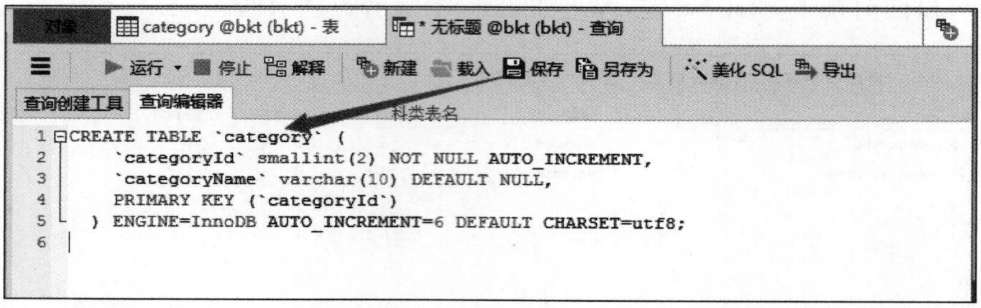

```
对象    category @bkt (bkt) - 表    * 无标题 @bkt (bkt) - 查询
   运行 ▸  停止  解释  新建  载入  保存  另存为  美化 SQL  导出
查询创建工具  查询编辑器
                                    科类表名
1  CREATE TABLE `category` (
2      `categoryId` smallint(2) NOT NULL AUTO_INCREMENT,
3      `categoryName` varchar(10) DEFAULT NULL,
4      PRIMARY KEY (`categoryId`)
5  ) ENGINE=InnoDB AUTO_INCREMENT=6 DEFAULT CHARSET=utf8;
6
```

图 5-86 编写 SQL 语句

CREATE TABLE 语句定义基本表,其基本格式如下所示:

CREATE TABLE <表名>(<列名><数据类型> [列级完整性约束条件]
    [,<列名><数据类型> [列级完整性约束条件]]…
    [,<表级完整性约束条件>])

建表的同时通常还可以定义与该表有关的完整性约束条件,这些完整性约束条件被存入系统的数据字典中,当用户操作表中数据时,由 RDBMS 自动检查该操作是否违背这些完整性约束条件。如果完整性约束条件涉及该表的多个属性"列",则必须定义在"表"级上;否则,既可以定义在"列"级,也可以定义在"表"级。

首先建立一张科类表(category),如程序清单 5-23 所示。

程序清单 5-23:bkt.sql

```
1   CREATE TABLE 'category' (
2      'categoryId' smallint(2) NOT NULL AUTO_INCREMENT,
3      'categoryName' varchar(10) DEFAULT NULL,
4      PRIMARY KEY ('categoryId')       /*列级完整性约束条件,categoryId 是主键*/
5   ) ENGINE=InnoDB AUTO_INCREMENT=6 DEFAULT CHARSET=utf8   /*编码格式*/
```

系统执行上面的 CREATE TABLE 语句后,就在数据库中建立一个新的空科类表,并且将有关科类表的定义及有关约束条件存放在数据字典中。MySQL 中创建相关表结构的 SQL 语句分别如下。

① 地区表(area)的建表语句如程序清单 5-24 所示。

程序清单 5-24:bkt.sql

```
1   DROP TABLE IF EXISTS 'area';                //如果存在 area 表,则先删除它
2   CREATE TABLE 'area' (                       //创建 area 表
3      'areaId' smallint(3) NOT NULL,
       //设置 areaId 字段,smallint 类型,长度是 3,不为空
4      'areaName' varchar(10) DEFAULT NULL,     //设置 areaName 字段,varchar 类
                                                //型,长度是 10,默认为空
5      PRIMARY KEY ('areaId')                   //将 areaId 设置为主键
6   ) ENGINE=InnoDB DEFAULT CHARSET=utf8;       //设置编码格式
```

第 5 章 数据库开发

② 批次表(batch)的建表语句如程序清单 5-25 所示。

程序清单 5-25：bkt.sql

```
1   DROP TABLE IF EXISTS 'batch';
2   CREATE TABLE 'batch' (
3   'batchId' smallint(2) NOT NULL AUTO_INCREMENT, //设置 batchId 自增
4      'batchName' varchar(10) DEFAULT NULL,
5      PRIMARY KEY ('batchId')
6   ) ENGINE=InnoDB AUTO_INCREMENT=6 DEFAULT CHARSET=utf8;
```

③ 类别表(category)的建表语句如程序清单 5-26 所示。

程序清单 5-26：bkt.sql

```
1   DROP TABLE IF EXISTS 'category';
2   CREATE TABLE 'category' (
3      'categoryId' smallint(2) NOT NULL AUTO_INCREMENT,
4      'categoryName' varchar(10) DEFAULT NULL,
5      PRIMARY KEY ('categoryId')
6   ) ENGINE=InnoDB AUTO_INCREMENT=6 DEFAULT CHARSET=utf8;
```

④ 省控线表(control_line)的建表语句如程序清单 5-27 所示。

程序清单 5-27：bkt.sql

```
1   DROP TABLE IF EXISTS 'control_line';
2   CREATE TABLE 'control_line' (
3      'controlId' int(10) NOT NULL,
4      'controlLine' smallint(5) DEFAULT NULL,
5      controlYear' smallint(5) DEFAULT NULL,
6      'categoryId' smallint(5) NOT NULL ,
7      'batchId' smallint(5) DEFAULT NULL,
8      'areaId' smallint(5) DEFAULT NULL,
9      PRIMARY KEY ('controlId'),
10     FOREIGN KEY ('areaId') REFERENCES 'area' ('areaId'),
                                         //设置外键,关联 area 表的 areaId
11     FOREIGN KEY ('batchId') REFERENCES 'batch' ('batchId'),
12     FOREIGN KEY ('categoryId') REFERENCES 'category' ('categoryId')
13  ) ENGINE=InnoDB DEFAULT CHARSET=utf8;
```

⑤ 学校表(school)的建表语句如程序清单 5-28 所示。

程序清单 5-28：bkt.sql

```
1   DROP TABLE IF EXISTS 'school';
2   CREATE TABLE 'school' (
```

```sql
3       'schoolId' smallint(6) NOT NULL,
4       'schoolName' varchar(50) DEFAULT NULL,
5       'schoolCode' int(10) DEFAULT NULL,
6       'areaId' smallint(3) NOT NULL,
7       'is985' bit(1) NOT NULL,
8       'is211' bit(1) NOT NULL,
9       'isMinistry' bit(1) NOT NULL,
10      PRIMARY KEY ('schoolId'),
11      FOREIGN KEY('areaId') REFERENCES 'area' ('areaId'),
12  ) ENGINE=InnoDB DEFAULT CHARSET=utf8;
```

⑥ 学校录取线表（school_recruit）的建表语句如程序清单 5-29 所示。

程序清单 5-29：bkt.sql

```sql
1   DROP TABLE IF EXISTS 'school_recruit';
2   CREATE TABLE 'school_recruit' (
3       'sRecruitId' int(10) NOT NULL,
4       'schoolId' smallint(6) DEFAULT NULL,
5       'sRecruitYear' smallint(5) DEFAULT NULL,
6       'batchId' smallint(5) DEFAULT NULL,
7       'categoryId' smallint(5) DEFAULT NULL,
8       'areaId' smallint(5) DEFAULT NULL,
9       'sRecruitAScore' smallint(5) DEFAULT NULL,
10      'sRecruitHScore' smallint(5) DEFAULT NULL,
11      'enterLine' smallint(5) DEFAULT NULL,
12      PRIMARY KEY ('sRecruitId'),
13      FOREIGN KEY ('areaId') REFERENCES 'area' ('areaId'),
14      FOREIGN KEY ('batchId') REFERENCES 'batch' ('batchId'),
15      FOREIGN KEY ('categoryId') REFERENCES 'category' ('categoryId'),
16      FOREIGN KEY ('schoolId') REFERENCES 'school' ('schoolId')
17  ) ENGINE=InnoDB DEFAULT CHARSET=utf8;
```

⑦ 专业表（major）的建表语句如程序清单 5-30 所示。

程序清单 5-30：bkt.sql

```sql
1   DROP TABLE IF EXISTS 'major';
2   CREATE TABLE 'major' (
3       'majorId' int(10) NOT NULL,
4       'majorName' varchar(50) DEFAULT NULL,
5       'majorCode' varchar(20) DEFAULT NULL,
6       PRIMARY KEY ('majorId'),
7   UNIQUE KEY 'majorCode' ('majorCode')          //设置唯一，不允许出现重复值
8   ) ENGINE=InnoDB DEFAULT CHARSET=utf8;
```

⑧ 专业录取线表(major_recruit)的建表语句如程序清单 5-31 所示。

程序清单 5-31：bkt.sql

```
1   DROP TABLE IF EXISTS 'major_recruit';
2   CREATE TABLE 'major_recruit' (
3       'majorRecruitId' int(10) NOT NULL,
4       'majorId' int(10) DEFAULT NULL,
5       'schoolId' smallint(6) DEFAULT NULL,
6       'year' smallint(5) DEFAULT NULL,
7       'mRecruitHScore' smallint(5) DEFAULT NULL,
8       'mRecruitAScore' smallint(5) DEFAULT NULL,
9       'categoryId' smallint(5) DEFAULT NULL,
10      'batchId' smallint(5) DEFAULT NULL,
11      'areaId' smallint(5) DEFAULT NULL,
12      PRIMARY KEY ('majorRecruitId'),
13      FOREIGN KEY ('majorId') REFERENCES 'major' ('majorId')
14      FOREIGN KEY ('areaId') REFERENCES 'area' ('areaId'),
15      FOREIGN KEY ('batchId') REFERENCES 'batch' ('batchId'),
16      FOREIGN KEY ('categoryId') REFERENCES 'category' ('categoryId')
17      FOREIGN KEY ('schoolId') REFERENCES 'school' ('schoolId')
18  ) ENGINE=InnoDB DEFAULT CHARSET=utf8;
```

⑨ 问题表(question)的建表语句如程序清单 5-32 所示。

程序清单 5-32：bkt.sql

```
1   DROP TABLE IF EXISTS 'question';
2   CREATE TABLE 'question' (
3       'questionId' int(10) NOT NULL AUTO_INCREMENT,
4       'questionContent' varchar(100) DEFAULT NULL,
5       'questionTime' varchar(20) DEFAULT NULL,
6       'answerContent' varchar(100) DEFAULT NULL,
7       'answerTime' varchar(100) DEFAULT NULL,
8       'questionStatus' varchar(20) DEFAULT NULL,
9       PRIMARY KEY ('questionId')
10  ) ENGINE=InnoDB AUTO_INCREMENT=5 DEFAULT CHARSET=utf8;
```

5.8.4 SQL 语句讲解

建好表之后，我们还需要往表中添加数据（当然也可以导入数据，后面章节会讲到），可以使用 SQL 语句来完成。另外，对数据的一些基本操作（如增、删、查、改）也可以用 SQL 语句实现，下面一一进行讲解。

同样以科类表为例，现在要向科类表中添加一条记录"categoryId＝1，categoryName ＝文史"。用 SQL 语句如何实现添加呢？

单击"查询"按钮,选中"新建查询"选项打开查询编辑器界面并输入查询语句,如图 5-87 所示。

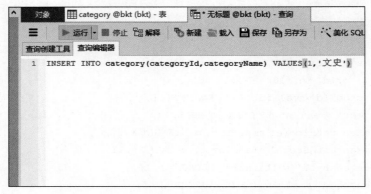

图 5-87　添加数据

然后单击"运行"按钮便可看到科类表中已增加一行记录。这里是使用 INSERT INTO 语句,category 是表名,括号里的 categoryId 和 categoryName 是表中的列名,VALUES 后面跟的是具体要插入的内容。

那如果要删除这一行数据呢?大家可以注意到设计表时 categoryId 是主键,也就是唯一值,我们可以根据 id 删除记录。删除 categoryId 为 1 的记录的语句如下:

DELETE FROM category WHERE categoryId=1;

如果我们在刚开始插入数据时不小心插入错了,现在要更改怎么办?同样也可以 id 为依据进行更新操作。把"文史"更改成"理工"的语句如下:

UPDATE category SET categoryName='理工' WHERE categoryId='1';

讲完增、删、改,终于到了最后一种操作。由于 SQL 查询涉及的内容实在太多,因此这里只是作简单介绍。如果你对 SQL 语言非常感兴趣,可以找一本专门介绍 SQL 的书进行学习。

查询科类表中的全部内容的语句如下:

SELECT * FROM category

等价于

SELECT categoryId,categoryName FROM category;

查询科类表中 categoryId 为 1 的详细记录的语句如下:

SELECT * FROM category WHERE categoryId=1;

查询科类表中科类名称是文史的那条记录的科类 id 的语句如下:

SELECT categoryId FROM category WHERE categoryName='文史';

以上 SQL 语句都是最基础的。因为在后面服务器章节完成案例时要连接数据库,也会用到 SQL 语句操作,所以为便于理解和加深印象,在此不妨先提前学习。具体 SQL 语句操作如下。

查询各省不限批次、不限科类、不限年份的省控线的 SQL 语句:

```
SELECT                                  //开始查询
    controlYear,                        //年份
    controlLine,                        //省控线
    areaName,                           //地区名称
    batchName,                          //批次名称
    categoryName                        //科类名称
    FROM                                //从哪里查
    control_line l,                     //省控线表,l 作为这张表的标记
    area a,                             //地区表,a 作为这张表的标记
    batch b,                            //批次表,b 作为这张表的标记
    category c                          //科类表,c 作为这张表的标记
WHERE                                   //"在哪里"表示条件约束
    l.areaId = a.areaId                 //当省控线表的 id 等于地区表的 id 时
AND l.batchId = b.batchId               //当省控线表的 id 等于批次表的 id 时
AND l.categoryId = c.categoryId         //省控线表的 id 等于科类表的 id 时
```

在查询编辑器中运行以上 SQL 语句,可看到查询结果,以下仅列出部分查询结果,如图 5-88 所示。

controlYear	controlLine	areaName	batchName	categoryName
2010	400	上海	二本	理工
2010	404	上海	二本	文史
2010	465	上海	一本	理工
2010	464	上海	一本	文史
2011	393	上海	二本	理工
2011	412	上海	二本	文史
2011	462	上海	一本	理工
2011	468	上海	一本	文史
2012	342	上海	二本	理工
2012	379	上海	二本	文史
2012	151	上海	专科	理工

图 5-88 省控线查询结果

查询专业分数线与学校等信息的 SQL 语句如下所示。

```
SELECT
    majorRecruitId,
    mr.YEAR,
    majorName,
    schoolName,
    batchName,
    categoryName,
    areaName,
    mRecruitHScore,
```

```
        mRecruitAScore
FROM
    major m,
    school s,
    area a,
    batch b,
    category c,
    major_recruit mr
WHERE
    mr.majorId = m.majorId
    AND mr.schoolId = s.schoolId
    AND mr.areaId = a.areaId
    AND mr.batchId = b.batchId
    AND mr.categoryId = c.categoryId
ORDER BY                            //根据批次 id 从小到大排序
    mr.batchId
```

在查询编辑器中运行以上 SQL 语句,可看到查询结果,以下仅列出部分查询结果,如图 5-89 所示。

majorRecruitId	YEAR	majorName	schoolName	batchName	categoryName	areaName	mRecruitHScore	mRecruitAScore
18293	2012	资源环境科学	北京师范大学	一本	理工	北京	615	611
101571	2011	新能源科学与工程	华北电力大学	一本	理工	江西	600	597
351167	2012	医学检验	上海交通大学医学院	一本	理工	上海	484	476
85065	2011	工商管理	中央民族大学	一本	文史	北京	588	573
304867	2012	宝石及材料工艺学	北京.中国地质大学	一本	理工	上海	439	436
764607	2013	测绘工程	东华大学	一本	理工	江西	562	561
49470	2012	英语	中国政法大学	一本	文史	上海	491	491
236520	2012	信息管理与信息系统	中国石油大学(北京校区)	一本	理工	北京	540	540
11956	2010	英语	上海大学	一本	文史	上海	492	477
789658	2010	会计学	北京语言大学	一本	理工	江西	560	560
762145	2011	机械	上海海事大学	一本	理工	江西	563	559
42109	2011	市场营销	华北电力大学	一本	理工	北京	568	568
169807	2013	法学	华东政法大学	一本	理工	上海	481	447

图 5-89　专业录取线信息查询结果

5.8.5　将 Excel 表导入数据库

还是以科类表为例,将数据保存在一张 Excel 表中,再将这张 Excel 表中的数据导入数据库的科类表。新建一个空的 Excel 文件用于保存科类信息,表中内容如图 5-90 所示。

然后选中数据库中的 category 表,右键选中"导入向导"选项,如图 5-91 所示。

进入导入向导页面,在"导入类型"这一栏选中"Excel 文件(2007 或以上版本)",然后单击"下一步"按钮,如图 5-92 所示。

	A	B	C
1	科类ID	科类名称	
2	1	文史	
3	2	理工	
4	3	综合	
5	4	艺术类	
6	5	体育类	
7			
8			
9			
10			

图 5-90　科类表(Excel 文件)

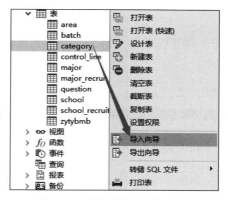

图 5-91　选择"导入向导"选项

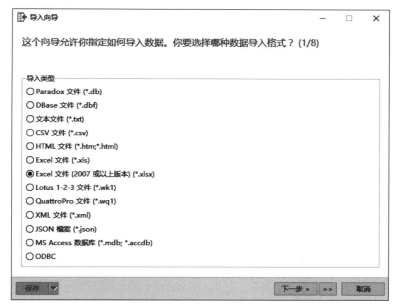

图 5-92　选中导入类型

"导入从"这一栏选择要导入的 Excel 表的存放地址，我们选择存到桌面，并且选中 Sheet1（Excel 表的内容是放在 Sheet1 中）的，然后单击"下一步"按钮（如图 5-93 所示）。

进入为源定义一些附加选项的页面，一般用默认选项就行，直接单击"下一步"按钮，如图 5-94 所示。

进入选择目标表页面，这里我们是把桌面上科类表（Excel）中 Sheet1 页面的内容导入数据库的 category 表，也就是我们的目标表，因此不用再新建一张表。故不用勾选，然后单击"下一步"按钮即可，如图 5-95 所示。

在源栏位选择与目标栏位相对应的列名，数据库表中的 categoryId 对应 Excel 表中的科类 ID，categoryName 对应科类名称，然后单击"下一步"按钮，如图 5-96 所示。

选择"添加"选项，然后单击"下一步"按钮，如图 5-97 所示。

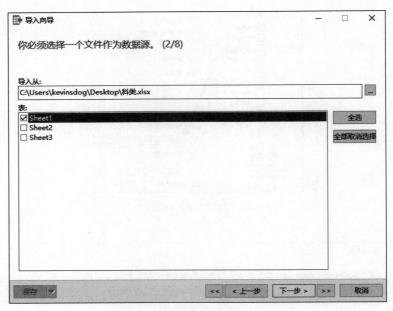

图 5-93 选择文件作为数据源

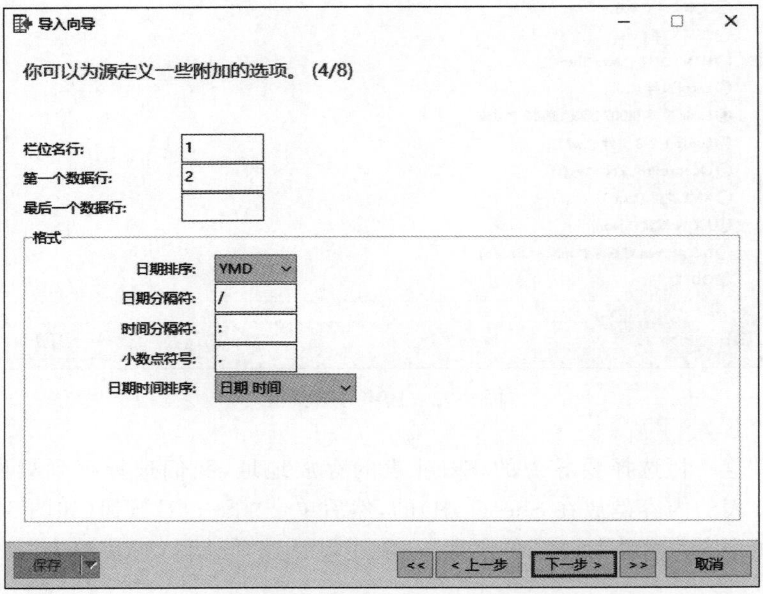

图 5-94 定义附加选项

然后单击"开始"按钮,完成数据导入,如图 5-98 所示。

至此,将整个 Excel 表导入数据库表中的操作就完成了,看上去步骤有点多,但每一步都很好理解,按照指示一步步进行就可。当然,除了导入 Excel 数据,我们还可以导入文本文档(.txt)格式类型文件,导入步骤与上面大致相同,在此不作详细介绍。但要注意

图 5-95　选择目标表

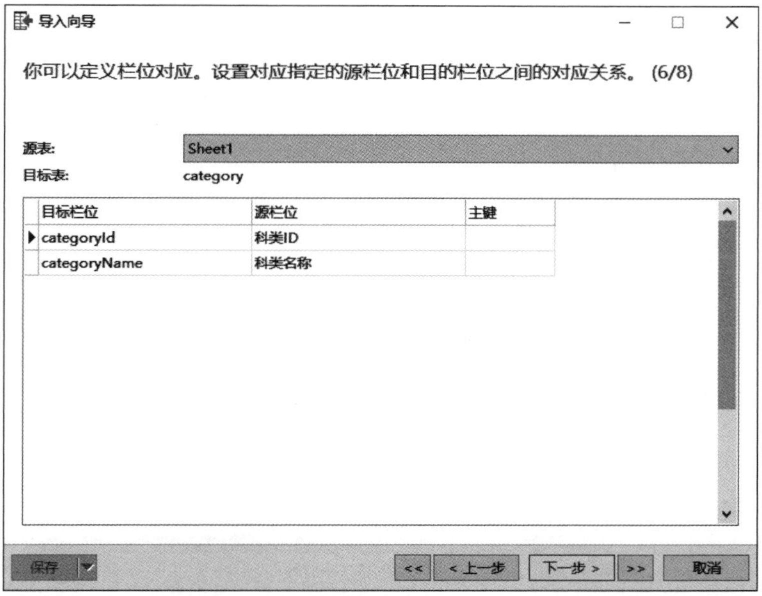

图 5-96　设置栏位对应关系

的一点是，文档中内容的格式要正确，每个栏位（对应数据库表中的列名）之间都要空一格（如图 5-99 所示）。

5.8.6　将 SQL 文件导入数据库

除了将 Excel 文件或文本文档文件导入数据库中，我们还可以直接将整个数据的

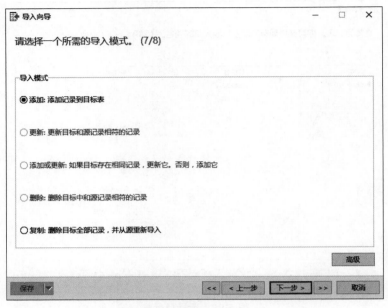

图 5-97 添加记录到目标表

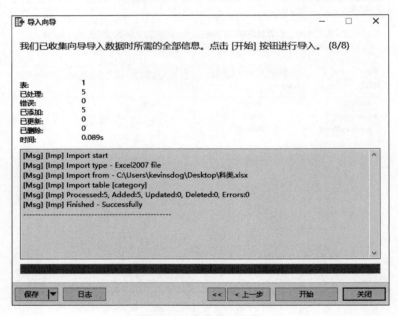

图 5-98 完成导入向导

SQL 文件导入。从提供的资源网址中下载 bkt.sql 文件,打开 MySQL,创建一个数据库,名为 bkt,指定编码方式为 utf-8,如图 5-100 所示。

创建完成后,选中 bkt 数据库,右键选择输入下的 SQL 文件,如图 5-101 所示。

第 5 章　数据库开发

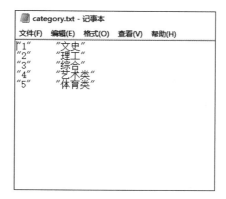

图 5-99　文本文档格式

图 5-100　创建数据库　　　　图 5-101　导入 SQL 文件

这时将弹出一个文件选择对话框，选择你下载的 bkt.sql 文件存放的路径，即进入导入"豹考通"数据的界面，如图 5-102 所示。

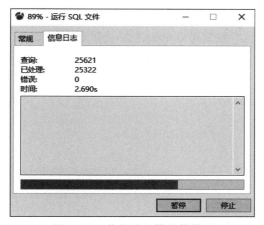

图 5-102　执行导入操作的界面

导入成功后刷新数据库,bkt 数据库下将会多出 9 张表,如图 5-103 所示。

图 5-103　导入成功后的 bkt 数据库

5.9　本章小结

本章主要介绍两种数据库(SQLite 和 MySQL)。SQLite 是轻量级数据库,可以安装在 Windows 操作系统,也可以在 Android 本地可以使用的数据库。本章讲解 SQLite 在 Windows 操作系统中下载和安装过程,并在 SQLite 中创建数据库、创建表和导入数据。讲解了使用 Java 代码在 Android Studio 中实现 SQLite 创建数据库,获取数据库和查询数据。项目中 Spinner 中数据获取方式由从类数组中获取变更为从本书提供的 SQLite 数据库获取。本章还重点讲解了 MySQL 数据库的相关知识以及它的特点和安装,可以通过图形化工具 Navicat for MySQL 进行访问。在一个项目中,数据库设计起到很关键的作用,本章围绕"豹考通"项目进行数据库设计,然后介绍了对应的数据库创建、表创建、数据的导入以及 SQL 语句的使用,为后面的服务器开发做好数据准备。

5.10　课后练习

1. 请简要概述 SQLite 的优缺点。
2. 安装好 Navicat for SQLite 软件并打开,对照本章 5.3 节内容,了解软件的基本操作,创建 bkt 本地数据库。
3. 简述 SQLite 数据库的简要特性及其常用的几个类属性。
4. 为建好的"豹考通"本地数据库(bkt.sqlite)实现 area 表和 school 表的数据替换。
5. 使用 Navicat for MySQL 软件进行建库建表操作。

第 6 章

Java Web 服务器端设计

6.1 本章简介

由于手机的计算能力和存储能力都比较有限,因此它通常作为移动终端使用,具体的数据处理则交给网络服务器来完成。它主要的优势在于携带方便,随时随地地访问网络和获取数据,因此利用手机与网络资源进行交互非常重要。网络提供资源需要服务器作为数据支撑,因此 Android 应用开发需要服务器开发配合,才能给用户更好的服务。

在 Android 客户端中,省控线查询模块、录取线查询模块、问题模块需要从服务获取数据,因此需要服务器配合开发。本章主要讲解"豹考通"服务器开发,其中会讲解开发环境搭建,Java Web 相关技术包括 DAO 设计模式、JavaBean、Servlet、JDBC 等。我们先以省控线查询作为代表实现查询服务,然后实现问题模块。

6.2 服务器开发背景知识

"工欲善其事,必先利其器",我们学习服务端的开发,必须先熟悉用于开发的工具。现在比较流行的开发工具有 Eclipse、MyEclipse 和 WebBuilder 等。这里推荐大家使用 MyEclipse 软件进行服务端的开发,因为其整合了专门用于开发 Web 应用的各类框架和功能,例如 Spring 框架、Hibernate 框架等。本书使用的版本是 MyEclipse 2015。

6.2.1 JSP 简介

JSP(Java Server Pages)是一种动态网页开发技术,这项技术是在 Servlet 的基础之上开发的,继承了 Java Servlet 所有的优秀功能。使用 JSP 可以十分高效地创建和开发具有安全性高、可跨平台特点的 Web 应用程序。

6.2.2 Tomcat 服务器

Tomcat 服务器是一个免费的开放源代码的 Web 应用服务器,属于轻量级应用服务器,在中小型系统和并发访问用户不是很多的场合下被普遍使用,是开发和调试 JSP 程序的首选。对于一个初学者来说,如果在一台机器上配置好 Apache 服务器,则可利用它响应 HTML 页面的访问请求。实际上,Tomcat 是 Apache 服务器的扩展,但它是独立运行的,因此当运行 Tomcat 时,它实际上作为一个与 Apache 独立的进程单独运行。也就是说,如果将你的项目放入 Tomcat 服务器中,你就可以去访问它。

我们所使用的 MyEclipse 2015 开发工具已经集成了 Tomcat 7.0 版本。但是,出于学习的目的，还是需要掌握如何配置和搭建 Tomcat 服务器。

1. 下载 Tomcat

首先，我们进入 Apache 官网(http://tomcat.apache.org/)下载 Tomcat，如图 6-1 所示。

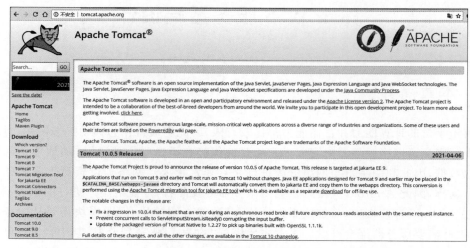

图 6-1　下载 Tomcat 服务器

找到官网左侧的 Download 栏目，选择你所需的 Tomcat 版本，这里我们选择 Tomcat 8，如图 6-2 所示。

单击 Tomcat 8 链接，选择系统对应版本进行下载，如图 6-3 所示。

2. 建立 tomcat 文件夹

在 MyEclipse 同级文件夹下建立一个 tomcat 文件夹，用于存放 Tomcat 服务器文件，如图 6-4 所示。

然后将下载的压缩包解压到 tomcat 文件夹下，如图 6-5 所示。　图 6-2　选择 Tomcat 8

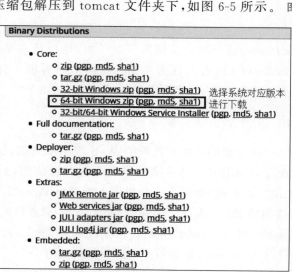

图 6-3　选择系统对应版本

图 6-4　新建 tomcat 文件夹

图 6-5　解压 Tomcat 安装包

3. 添加 Tomcat 服务器

接下来,打开 MyEclipse 2015,进入偏好设置界面,找到图 6-6 中箭头指向的地方。

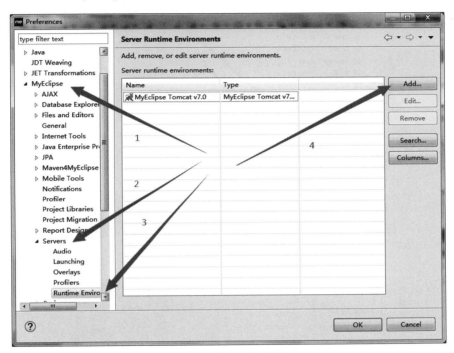

图 6-6　偏好设置界面

单击 Add 按钮进入图 6-7 所示的界面。

单击 Next 按钮进入图 6-8 所示的界面。

单击 Finish 按钮后回到 MyEclipse 主界面,在 Servers 选项区域空白处右击,如图 6-9 所示。

单击 Server 按钮,显示图 6-10 所示界面。

单击 Finish 按钮后即成功添加 Tomcat 8 服务器。

6.2.3　服务器与客户端交互

客户端与服务器交互根据 HTTP(hypertext transfer protocol)协议来进行,HTTP 协议是一种超文本传输协议,是计算机在网络中通信的一种规则。它也是一种无状态的协议,意思是指在 Web 浏览器(客户端)和 Web 服务器之间不需要建立持久的连接。整个过程就是由客户端向服务器端发送一个请求,然后 Web 服务器返回一个响应,之后连接关闭,在服务器端不保留连接信息。

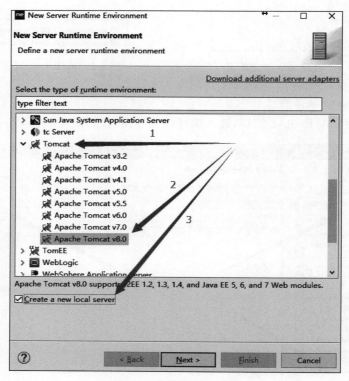

图 6-7　选择运行时环境

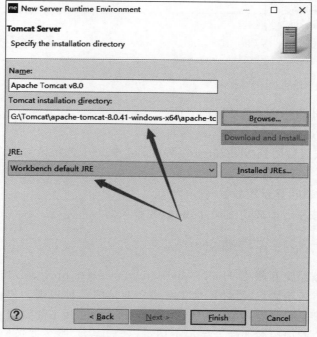

图 6-8　指定安装目录

图 6-9　显示右键菜单

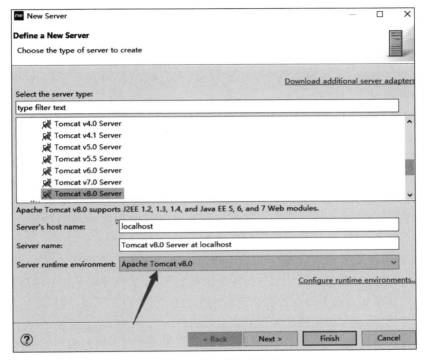

图 6-10　选择服务器类型

交互过程主要有以下几个步骤（见图 6-11）：

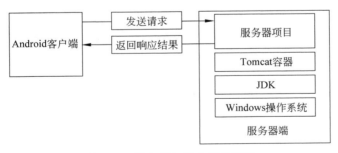

图 6-11　服务器与客户端交互图

(1) 建立连接。
(2) 客户端发送 HTTP 请求头。

(3) 服务器端响应生成结果"回发"。

(4) 服务器端关闭连接，客户端解析"回发"响应头。

6.3 了解 Java Web 技术

6.3.1 DAO 设计模式

DAO(Data Access Object)设计模式实际上是两个模式的组合，即 Data Accessor 模式和 Active Domain Object 模式。其中 Data Accessor 模式实现了数据访问和业务逻辑的分离，而 Active Domain Object 模式实现了业务数据的对象化封装，一般我们将这两个模式组合使用。DAO 设计模式属于 J2EE 数据层的操作，使用它可以简化大量代码，增强程序的可移植性。

DAO 模式实现了以下 4 个目标。

1. 数据存储逻辑的分离

通过对数据访问逻辑进行抽象，为上层机构提供抽象化的数据访问接口，业务层无须关心具体的数据操作。这样一方面避免了业务代码中混杂 JDBC 调用语句，使得业务实现更加清晰；另一方面，由于数据访问实现分离，也使得开发人员的专业划分成为可能。某些精通数据库操作技术的开发人员可以根据接口提供数据库访问的最优化实现，而精通业务的开发人员则可以抛开数据的烦琐细节，专注于业务逻辑编码。

2. 数据访问底层实现的分离

DAO 模式通过将数据访问计划分为抽象层和实现层，从而分离了数据使用和数据访问的实现细节。这意味着业务层与数据访问的底层细节无关，也就是说，我们可以在保持上层机构不变的情况下，通过切换底层实现来修改数据访问的具体机制。常见的一个例子就是，我们可以通过仅替换数据"访问层"实现，将我们的系统部署在不同的数据库平台之上。

3. 资源管理和调度的分离

在数据库操作中，资源的管理和调度是一个非常值得关注的主题。大多数系统的性能瓶颈往往并非集中于业务逻辑处理本身。在系统涉及的各种资源调度过程中，存在着最大的性能黑洞，而数据库作为业务系统中最重要的系统资源，自然也成为关注的焦点。DAO 模式将数据访问逻辑从业务逻辑中脱离开来，使得在数据访问层实现统一的资源调度成为可能，可以在保持上层系统不变的情况下大幅度提升系统性能。

4. 数据抽象

在直接基于 JDBC 调用的代码中，程序员面对的数据往往是原始的 RecordSet 数据集。诚然，这样的数据集可以提供足够的信息，但对于业务逻辑开发过程而言，如此琐碎和缺乏寓意的字段型数据实在令人厌倦。

DAO 模式通过对底层数据的封装，为业务层提供一个面向对象的接口，使得业务逻辑开发员可以面向业务中的实体进行编码。通过引入 DAO 模式，业务逻辑更加清晰，且富于形象性和描述性，这将为日后的维护带来极大的便利。试想，在业务层通过 Customer.getName 方法获得客户姓名，相对于直接通过 SQL 语句访问数据库表并从

ResultSet 中获得某个字符型字段而言,哪种方式更加易于业务逻辑的形象化和简洁化?

后面所有的程序都将使用 DAO 设计模式的思想编写,学习者也应该养成这样的编码习惯。

6.3.2 认识 Java Web 程序的目录结构

我们新建一个 Web Project 项目(图 6-12),这里对项目中的各项内容进行一个详细的说明。

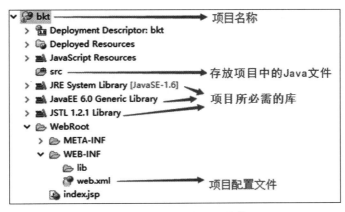

图 6-12 Web 程序目录结构

- src:用于存放 Java 源文件。
- WebRoot:用于存放 JSP、JS、CSS、图片等文件。
- lib:用于存放需要导入的库文件。
- web.xml:用于配置项目的欢迎页、Servlet、Filter 等。在 Web 3.0 下,已经可以不需要 web.xml 文件,可使用注解的方式进行配置。

6.4 Java Web 核心技术

6.4.1 JavaBean 技术

1. JavaBean 概述

JavaBean 组件就是用 Java 语言编写的组件,它就像一个封装好的容器,里面具有满足用户使用要求的功能。每个 JavaBean 实现一个特定的功能,通过合理地组织不同功能的 JavaBean,可以快速生成一个全新的应用程序。

2. JavaBean 技术介绍

使用 JavaBean 的最大优点在于它可以提高代码的重用性。一个成功的 JavaBean 需要符合"一次性编写,任何地方执行,任何地方重用"条件。所谓一次性编写,就是在组件重用时不需要重新编写,只需要根据需求修改和升级代码即可;任何地方执行,是指可以在任何平台上运行,由于 JavaBean 是基于 Java 语言编写的,所以它可以轻易地移植到各种平台之上;任何地方重用是指JavaBean组件能够被放在多种方案中使用,包括应用程序、其他组件、Web应用等。6.5节编写的数据库工具类就是一个JavaBean。

6.4.2 运行第一个Java Web程序

1. 创建Web项目

在MyEclipse的主界面中找到左上角的File菜单项并单击,在弹出的下拉菜单中单击New选项,再选择Web Project选项即可(图6-13和图6-14所示)。

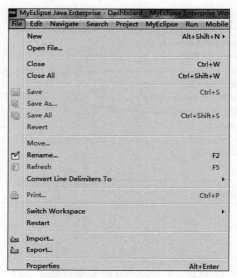

图6-13 选择File→New选项

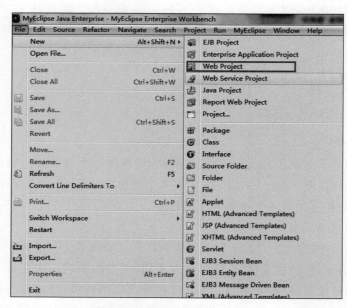

图6-14 选择Web Project选项

此时弹出图 6-15 所示的对话框。

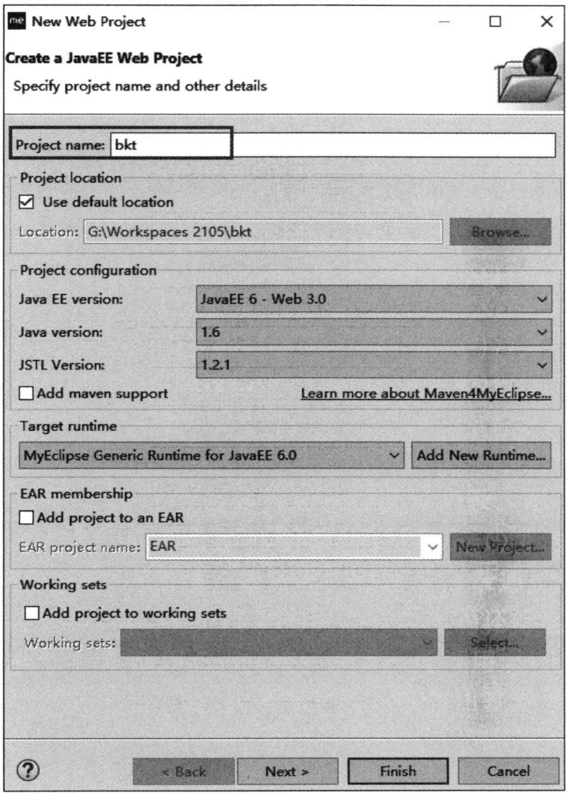

图 6-15　New Web Project 对话框

Project name 是项目名，这里取名为 bkt；Project location 是项目的存放地址，这里使用默认的地址；Java EE version 栏选择 JavaEE 6-Web 3.0。单击 Finish 按钮完成创建。

2. 将项目导入 Tomcat 服务器中

右击 Server 下的 MyEclispe Tomcat v7.0，选择 Add/Remove Deployments 选项，如图 6-16 所示。

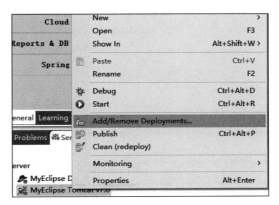

图 6-16　选择 Add/Remove Deployments 选项

选中我们所创建的项目 bkt，单击 Add 按钮，然后单击 Finish 按钮，如图 6-17 所示。

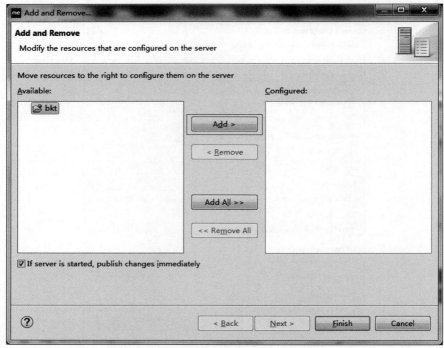

图 6-17　添加项目 bkt

3. 启动 Tomcat 服务器

右击 MyEclispe Tomcat v7.0，选择 Start 选项，启动 Tomcat 服务器，如图 6-18 所示。

4. 检测 Web 项目是否创建成功

打开浏览器，在地址栏输入 http://localhost:8080/bkt/（其中，localhost 代表 Tomcat 服务器的地址是本机，8080 代表 Tomcat 的默认端口号是 8080，后面紧跟项目名），就可以访问项目的首页。如果出现图 6-19 所示的结果，就说明项目已经成功运行。

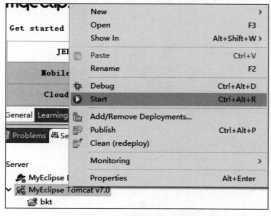

图 6-18　启动 Tomcat 服务器

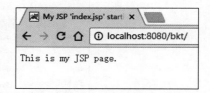

图 6-19　项目运行结果

6.4.3 Servlet 技术

Servlet 是一种独立于平台和协议的服务器端的 Java 技术,可以用来生成动态的 Web 页面。Servlet 具有良好的可移植性、强大的功能、高效率、低投资、安全性高的特点。

Servlet 通过创建一个框架来扩展服务器的能力,为 Web 上进行请求提供相应的服务。当客户端发送请求至服务器时,服务器可以将请求信息发送给 Servlet,并让 Servlet 建立起响应返回给客户端。当启动 Web 服务器或客户端第一次请求服务时,可以自动装入 Servlet,之后 Servlet 继续运行,直到其他客户端发出请求。Servlet 是一个 Java 类,Java 语言能够实现的功能,它基本上都可以实现(图形界面除外)。总的来说,Servlet 是运行在服务器端的程序,用于处理及响应客户端的请求。

在 Servlet 的整个生命周期中,其处理过程如图 6-20 所示。

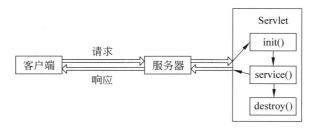

图 6-20 Servlet 的处理过程

图 6-20 所示各步骤的说明如下:

(1) 用户通过客户端浏览器请求服务器,服务器加载 Servlet 并创建一个 Servlet 实例。

(2) 容器调用 Servlet 的 init()方法。

(3) 容器调用 Servlet 的 service()方法,并将 HttpServletRequest 和 HttpServletResponse 对象传递给该方法,在 service()方法中处理用户请求。

(4) 在 Servlet 中处理完请求后,将结果返回给容器。

(5) 容器将结果返回给客户端进行显示。

(6) 当 Web 容器关闭时,调用 destroy()方法销毁 Servlet 实例。

6.4.4 HttpServletRequest 类

request 对象代表客户端的请求信息,主要用于接受通过 HTTP 协议传送到服务器的数据(包括头信息、系统信息、请求方式以及请求参数等)。request 对象的作用域为一次请求。

当客户端向服务器端发送请求时,服务器为本次请求创建 request 对象,并在调用 Servlet 的 service 方法时将该对象传递给 service 方法。request 对象中封装了客户端发送过来的所有请求数据。

request 对象获取请求参数的方法如表 6-1 所示。

表 6-1 request 对象的方法

方　　法	说　　明
String getParameter(String name)	获取指定名称的请求参数值,适用于单值的请求参数
String[] getParameterValues(String name)	获取指定名称的请求参数值,适用于多值的请求参数
Enumeration<String> getParameterNames()	获取所有的请求参数名称
Map<String,String[]> getParameterMap()	获取所有请求参数,其中参数名作为 Map 的 key,参数值作为 Map 的 value

6.4.5 HttpServletResponse 类

返回一个 Map 类型值,其中记录了前端请求参数和前端请求参数值的映射关系。response 对象代表的是对客户端的响应,主要是将服务器处理过的对象传回到客户端。response 对象也具有作用域,它只在 JSP 页面内有效,其常用方法如表 6-2 所示。

表 6-2 response 对象的常用方法

方　　法	说　　明
response.write()	将指定的字符串写到当前的 HTTP 输出
response.redirect()	使浏览器从当前网页转向其他网页

6.4.6 实战演练

在本节中,我们将通过实际的小项目来加深对 request 对象与 response 对象的理解。
(1) 首先新建一个名为 code0601 的 Web 项目。
(2) 在 src 文件夹下建立一个名为 drhelp.cn.servlet 的包,方法是右击 src 文件夹,选择 New 选项,再选择 Package 选项(见图 6-21)。
(3) 新建 Servlet 类。右击 drhelp.cn.servlet 包,操作如图 6-22 所示。
(4) 弹出图 6-23 所示的对话框,填写 Servlet 类名。
(5) 单击 Finish 按钮。进入创建好的 Servlet 类中,可以看到这个 Servlet 有 init()方法、destroy()方法、doGet()方法和 doPost()方法。init 和 destroy 方法前面已经介绍过。doGet 方法和 doPost 方法分别用来处理客户端提出的 Get 请求和 Post 请求。Get 请求和 Post 请求是指定了 method="Get"或者 method="Post"的表单时发出的请求。

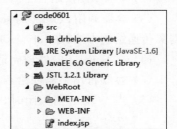

图 6-21 新建 drhelp.cn.servlet 包

我们在 doPost()方法中定义一个 String 类型对象 year,用于存放客户端传递过来的值,再通过 response 对象将 year 传递给前台,如程序清单 6-1 所示。

第 6 章　Java Web 服务器端设计

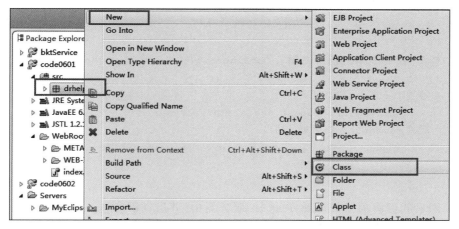

图 6-22　新建 Servlet 类

图 6-23　填写 Servlet 类名

程序清单 6-1：Code0601\src\drhelp\cn\servlet\Code0601Ser.java

```
1   @WebServlet("/Code0601Ser")
2   public class Code0601Ser extends HttpServlet {
3       private static final long serialVersionUID = 1L;
4       public Code0601Ser() {
5         super();
6         //TODO Auto-generated constructor stub
7       }
8       protected void doGet(HttpServletRequest request, HttpServletResponse
    response) throws ServletException, IOException {
9           //TODO Auto-generated method stub
10          this.doPost(request, response);
11      }
12      protected void doPost(HttpServletRequest request, HttpServletResponse
```

219

```
              response) throws ServletException, IOException {
13    / TODO Auto-generated method stub
14    /*
      测试方法是在浏览器地址栏输入 localhost:8080/code0601/Code0601Ser?year=2011
15    */
16    //通过 request 获取字段内容
17        String year=request.getParameter("year");
18    //通过 response 进行数据的输出
19        response.getOutputStream().write(year.getBytes("utf-8"));
20    }
21  }
```

（6）随后我们将项目部署到 Tomcat 服务器中，并启动服务器。打开浏览器，在地址栏输入 http://localhost:8080/code0601/Code0601Ser?year=2011。

（7）结果如图 6-24 所示。

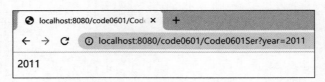

图 6-24　程序运行结果图

我们通过 request 对象的 getParameter("year")方法来获取 year 的值，并存入新建变量 year，再通过 response 对象的方法将获取的 year 变量传递给前台显示。

6.5　设计 App 服务器数据库工具类

自本节起，我们将开始正式编写"豹考通"后台服务器项目。本节将创建一个 DBconn 类，这个类的作用是连接数据库。

6.5.1　JDBC 技术

1. JDBC 是什么

JDBC 是一种可用于执行 SQL 语句的 Java API，它由一些 Java 语言编写的类和界面组成。JDBC 为数据库应用开发人员、数据库前台工具开发人员提供了一种标准的应用程序设计接口，使开发人员可以用纯 Java 语言编写完整的数据库应用程序。

2. JDBC 的作用

JDBC 的作用可简单归纳为如下 3 点。

- 与数据库建立连接。
- 向数据库发送 SQL 语句。
- 处理数据库返回的结果。

3. JDBC 驱动

我们要使用 JDBC 进行数据库操作，就必须在项目中导入 JDBC 驱动（JDBC 驱动可从网上下载）。导入步骤也十分简单，只需要把驱动程序包复制粘贴到 code0602\WebRoot\WEB-INF\lib 文件夹下，当 Web App Libraries 中出现此驱动时，则说明成功导入（见图 6-25）。

图 6-25　添加 JDBC 驱动

6.5.2　数据库连接类的实现

（1）新建一个名为 Code0602 的项目，随后在 src 下新建两个包，分别为 drhelp.cn.util（用来存放数据库连接类 DBConn.java）和 drhelp.cn.test（用来存放测试是否能够成功连接数据库的类 Test.java）。

（2）在 drhelp.cn.util 下新建 DBConn 类，如图 6-26 所示。

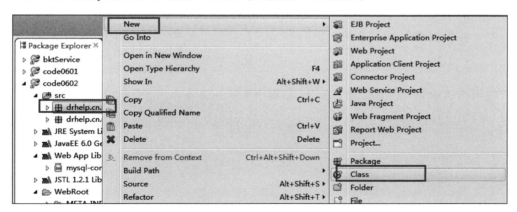

图 6-26　新建 DBConn 类

（3）在弹出的对话框的 Name 栏中填写类名 DBConn，单击 Finish 按钮（图 6-27 所示）。

（4）上一步骤已成功创建好 DBConn 类，接下来编写代码，实现连接数据库的功能。

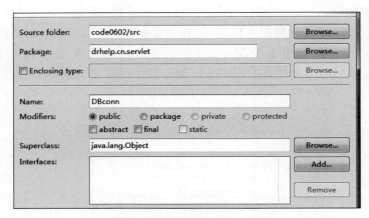

图 6-27　填写类名 DBconn

这个 DBConn 类需要以下 5 个变量：
- 连接数据库的 URL。
- 数据库驱动。
- 数据库用户名。
- 数据库连接密码。
- 创建数据库连接方法。

在 DBConn 类中定义程序清单 6-2 所示的变量。

程序清单 6-2：Code0602\src\drhelp\cn\servlet\DBconn.java

```
1    private String url;                    //dbc:mysql://
2    private String serverName;             //数据库地址,如 localhost
3    private String portNumber;             //数据库端口 3306
4    private String databaseName;           //数据库名
5    private String username;               //数据库用户名
6    private String password;               //数据库连接密码
```

使用构造函数对变量进行初始化,如程序清单 6-3 所示。

程序清单 6-3：Code0602\src\drhelp\cn\servlet\DBconn.java

```
1    public DBconn(){
2      url="jdbc:mysql://";
3      serverName="localhost";
4      portNumber="3306";
5      databaseName="bktserver?characterEncoding=utf-8";
6      username="root";
7      password="8888";
8    }
```

（5）设置好各个变量的参数后,需要将这些变量合成一个 URL。我们单独使用返回类型为 String 的方法 getConnnectUrl()来合成 URL,代码如程序清单 6-4 所示。

程序清单 6-4:Code0602\src\drhelp\cn\servlet\DBconn.java

```
1   public String getConnnectUrl(){
2       return url+serverName+":"+portNumber+"/"+databaseName;
3   }
```

(6)接下来开始编写一个连接数据库的方法,这个方法返回的类型是一个数据库连接对象 Connection,这里我们将这个方法命名为 getConnection(),实现代码如程序清单 6-5 所示。

程序清单 6-5:Code0602\src\drhelp\cn\servlet\DBconn.java

```
1   public Connection getConnection(){
2       Connection con=null;
3       try{
4           Class.forName("com.mysql.jdbc.Driver");
5           con=DriverManager.getConnection(getConnnectUrl(),username, password);
6           System.out.println("连接数据库成功!");
7       }catch (Exception e) {
8           e.printStackTrace();
9           System.out.println("连接数据库失败:"+e.getMessage());
10      }
11      return con;
12  }
```

我们通过 Class 对象的 forName 方法加载数据库驱动,再通过 DriverManager 对象的 getConnection 方法获取数据库连接。

写好连接方法后,可以运行这个方法,查看能否成功连接至数据库,如果连接失败,则系统会打印异常信息。我们在 drhelp.cn.test 包中新建一个 Test.java 文件。

Test.java 文件的代码如程序清单 6-6 所示。

程序清单 6-6:code0602\src\drhelp\cn\test\Test.java

```
1   package drhelp.cn.test;
2   import drhelp.cn.util.DBconn;
3   public class Test {
4       public static void main(String[] args) {
5           DBconn db=new DBconn();
6           db.getConnection();
7       }
8   }
```

运行此测试文件,如果结果如图 6-28 所示,则说明成功连接数据库(测试前确保数据库开启)。

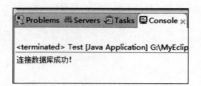

图 6-28 测试结果

DBConn 类的代码如程序清单 6-7 所示。

程序清单 6-7：Code0602\src\drhelp\cn\util\DBConn.java

```
1   package drhelp.cn.util;
2   import java.sql.Connection;
3   import java.sql.DriverManager;
4   public class DBconn {
5       private String url;                    //dbc:mysql://
6       private String serverName;             //数据库地址,如 localhost
7       private String portNumber;             //数据库端口 3306
8       private String databaseName;           //数据库名
9       private String username;               //数据库用户名
10      private String password;               //数据库连接密码
11      public DBconn(){
12         url="jdbc:mysql://";
13         serverName="localhost";
14         portNumber="3306";
15         databaseName="bktserver?characterEncoding=utf-8";
16         username="root";
17         password="8888";
18      }
19      //jdbc:mysql://localhost:3306/
20      public String getConnnectUrl(){
21         return  url+serverName+":"+portNumber+"/"+databaseName;
22      }
23      public Connection getConnection(){
24         Connection con=null;
25         try{
26            Class.forName("com.mysql.jdbc.Driver");
27            con=DriverManager.getConnection(getConnnectUrl(), username, password);
28            System.out.println("连接数据库成功!");
29         }catch (Exception e) {
30            e.printStackTrace();
31            System.out.println("连接数据库失败:"+e.getMessage());
32         }
33         return con;
```

```
34      }
35  }
```

6.6 设计 App 服务器业务逻辑类

上一节我们只是单纯学习了如何连接数据库，但是无法从数据库中获取数据。本节将实现从数据库中获取数据。

6.6.1 建立实体类

在前面的章节中，我们学习了如何使用本地数据查询省控线、学校录取线和专业录取线。因此，第一步是需要确定"查询省控线""查询学校录取线"和"查询专业录取线"三个模块各需要将哪些数据查询出来。确定好所需的数据字段后，第二步再建立对应的 JavaBean 来存放所查询的数据。

我们先学习如何查询省控线。通过前面章节学习如何使用 MySQL 数据库，我们可以很快确定"查询省控线"模块需要在数据库中查询省控线、省份名、批次名、类别名和年份 5 个变量的值。

新建名为 bktService 的 Web 项目，或者在上一节的项目中接着做。bktService 项目目录结构如图 6-29 所示。

其中 drhelp.cn.test 包中的 Test.java 文件和 drhelp.cn.util 包中的 DBconn.java 文件就是我们上一节中所学的内容。

建立好图 6-29 所示的项目，drhelp.cn.bean 包下面存放所有的实体类，drhelp.cn.impl 包下面存放所有的数据库操作类。我们首先在 drhelp.cn.bean 包下建立一个实体类 ControlLine。在 ControlLine 类中需要有 5 个成员变量（省控线、省份名、批次名、类别名和年份）（见程序清单 6-8）。

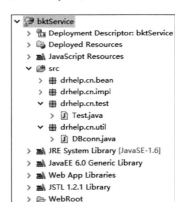

图 6-29 bktService 项目目录结构

程序清单 6-8：code0603\bktService\src\cn\jxufe\iet\bean\ControlLine.java

```
1   package drhelp.cn.bean;
2   import java.io.Serializable;
3   @SuppressWarnings("serial")
4   public class ControlLine implements Serializable {      //省控线封装类
5       private int controlLine;                            //省控线
6       private String areaName;                            //省份名
7       private String batchName;                           //批次名
8       private String categoryName;                        //类别名
9       private int controlYear;                            //年份
```

```
10      public int getControlLine() {
11      return controlLine;
12      }
13      public void setControlLine(int controlLine) {
14      this.controlLine = controlLine;
15      }
16
17      public String getAreaName() {
18      return areaName;
19      }
20      public void setAreaName(String areaName) {
21      this.areaName = areaName;
22      }
23      public String getBatchName() {
24      return batchName;
25      }
26      public void setBatchName(String batchName) {
27      this.batchName = batchName;
28      }
29      public String getCategoryName() {
30      return categoryName;
31      }
32      public void setCategoryName(String categoryName) {
33      this.categoryName = categoryName;
34      }
35      public int getControlYear() {
36      return controlYear;
37      }
38      public void setControlYear(int controlYear) {
39      this.controlYear = controlYear;
40      }
41      @Override
42      public String toString() {
44  return "ControlLine [cotrolLine=" + controlLine + ", areaName="
44  + areaName + ", batchName=" + batchName + ", categoryName="
45      + categoryName + ", controlYear=" + controlYear + "]";
46      }
47
48  }
```

由代码可知,我们重写了父类的 toString 方法来返回省控线信息。这是为了返回 JSON 格式,方便客户端解析数据。

6.6.2 数据库操作类的实现

要查询省控线,我们的 App 有如下 4 种情况:

- 不限批次、不限类别查询。
- 不限批次、限类别查询。
- 不限类别、限批次查询。
- 限批次、限类别查询。

因此，我们在进行程序设计时，需要分别处理这 4 种情况。查询省控线时，需要从 Android 客户端获取省控线年份、地区编号、批次编号和类别编号 4 个变量来进行数据库查询。我们知道，省控线年份和地区编号变量是肯定有值的，批次和类别是不确定的变量，可能为空或者为 0。那么如何通过这两个变量来处理上述 4 种情况呢？我们可以对这两个变量进行一个处理，当传值为空时，我们将它赋值为 −1。如果这两个变量为空或者为 0，则批次或类别不限，这个处理过程的实现在后面章节讲解。这里，我们只需要使用处理后的变量即可。

首先在 drhelp.cn.impl 包下建立一个名为 ControlLineImpl 的专门用于查询省控线的数据库操作类，如图 6-30 所示。

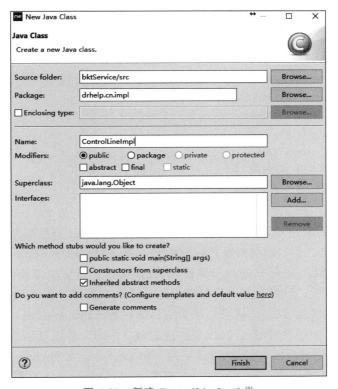

图 6-30　新建 ControlLineImpl 类

新建好 ControlLineImpl 类后，我们先写一个获取省控线的方法 getControlLine，这个方法有 4 个参数，分别为省控线年份 controlYear、地区编号 areaId、类别编号 categoryId 和批次编号 batchId，返回 ControlLine 实体类的 List 对象。代码如程序清单 6-9 所示。

程序清单 6-9：Code0603\bktService\src\cn\jxufe\iet\impl\ControlLineImpl.java

```
1    public List < ControlLine > getControLine (int controlYear, int areaId, int
     categoryId, int batchId ){
2        List<ControlLine> controlLine=new ArrayList<ControlLine>();
3        return controlLine;
4    }
```

为什么是 List 对象而不是 ControlLine 对象呢？这是因为省控线不一定只有一条数据，而 ControlLine 对象显然是不能满足条件的，所以需要使用 List 对象来存放查询的结果。

对数据库进行查询操作，就需要先获取数据库连接。因为我们已经在上一节中学习了连接类，所以只需要调用该类的方法就行。代码如程序清单 6-10 所示。

程序清单 6-10：Code0603\bktService\src\cn\jxufe\iet\impl\ControlLineImpl.java

```
1    public List < ControlLine > getControLine (int controlYear, int areaId, int
     categoryId, int batchId ){
2        List<ControlLine> controlLine=new ArrayList<ControlLine>();
3        Connection con=null;
4        ResultSet rs=null;
5        PreparedStatement pre=null;
6        DBconn db=new DBconn();
7        con=db.getConnection();
8        return controlLine;
9    }
```

首先我们定义 3 个分别为 Connection 类型、ResultSet 类型和 PreparedStatement 类型的变量。Connection 对象是 java.sql 包下的一个类，主要作用是获得数据库的连接，能够创建一个 PreparedStatement 对象，将参数化的 SQL 语句发送到数据库。PreparedStatement 对象是预编译的，它把 SQL 语句先进行预编译，然后再执行。简单地说，所有 SQL 语句运行前必然有个解析过程，就好像 Java 代码的编译一样。如果同样的语句运行多次，当然希望它是一次编译多次运行的，这样可以减少编译的过程。PreparedStatement 就是用于这一目的。它其实返回一个编译后的标识。通过这个标识，系统可以直接找到编译过的 SQL 来运行，不必把每次的 SQL 都编译一遍。ResultSet 是数据查询结果返回的一种结果集，可以说结果集是一个存储查询结果的对象，但是结果集并不只具有存储的功能，它同时还具有操纵数据的功能，可能完成对数据的更新等。我们定义 ResultSet 对象是为了保存我们所查询的结果。随后，我们定义一个 DBConn 对象，调用该对象的 getConnection() 方法来获取数据库连接。

获取数据库连接之后，我们就可以对数据库进行查询操作。前面章节已经详细讲解了查询省控线的 SQL 语句，但是查询省控线有 4 种情况，因此我们需要动态地对 SQL 进行处理，然后再执行 SQL 语句。前面说到，我们通过类别编号和批次编号的值是 0 或 −1

来判断是否需要考虑类别和批次。动态处理过程如程序清单 6-11 所示。

程序清单 6-11：Code0603\bktService\src\cn\jxufe\iet\impl\ControlLineImpl.java

```
1   String sql =" select controlYear, controlLine, areaName, batchName,
    categoryName from control_line l,area a,batch b,category c where l.areaId=
    a.areaId and l.batchId = b.batchId and l.categoryId = c.categoryId and
    controlYear=? and l.areaId=?";
2   if(batchId==0||batchId==-1){                    //批次不限
3       if(categoryId==0||categoryId==-1){          //类别不限
4           sql+=" order by l.categoryId";          //查询结果按照类别排序
5       try {
6           pre=con.prepareStatement(sql);
7           pre.setInt(1, controlYear);
8           pre.setInt(2, areaId);
9           rs=pre.executeQuery();
10      } catch (SQLException e) {
11          printStackTrace();
12      }
13      }else{
14          sql+=" and l.categoryId=? order by l.categoryId";
15      try {
16          pre=con.prepareStatement(sql);
17          pre.setInt(1, controlYear);
18          pre.setInt(2, areaId);
19          pre.setInt(3, categoryId);
20          rs=pre.executeQuery();
21      } catch (SQLException e) {
22          printStackTrace();
23      }
24      }
25  }else{
26      sql+=" and l.batchId=? order by l.categoryId";
27      try {
28          pre=con.prepareStatement(sql);
29          pre.setInt(1, controlYear);
30          pre.setInt(2, areaId);
31          pre.setInt(3, batchId);
32          rs=pre.executeQuery();
33      } catch (SQLException e) {
34          e.printStackTrace();
35      }
36      if(categoryId==0||categoryId==-1){          //类别不限
37      }else{
```

```
38      sql+=" and l.categoryId=? order by l.categoryId";
39    try {
40      pre=con.prepareStatement(sql);
41      pre.setInt(1, controlYear);
42      pre.setInt(2, areaId);
44      pre.setInt(3, batchId);
44      pre.setInt(4, categoryId);
45      rs=pre.executeQuery();
46    } catch (SQLException e) {
47      e.printStackTrace();
48        }
49    }
50  }//根据请求参数创建对应的SQL语句
```

先是编写基础的 SQL 语句,我们要传入的变量用?代替。4 种情况的处理思想如下: 首先我们考虑批次是否为 0 或 −1,为 0 或 −1 则代表查询省控线不限批次。然后在之前的情况下再分别考虑类别是否为 0 或 −1,即查询省控线是否考虑类别。

在这里,我们先使用 if 语句判断批次编号 batchId 是否为 0 或 −1,如果 if 为真,再使用 if 语句来判断类别编号 categoryId 的值是否为 0 或 −1(程序清单 6-12)。

程序清单 6-12：Code0603\bktService\src\cn\jxufe\iet\impl\ControlLineImpl.java

```
1   if(categoryId==0||categoryId==-1){        //类别不限
2     sql+=" order by l.categoryId";          //查询结果按照类别排序
3     pre=con.prepareStatement(sql);
4     pre.setInt(1, controlYear);
5     pre.setInt(2, areaId);
6     rs=pre.executeQuery();
7   }else{
8     sql+=" and l.categoryId=? order by l.categoryId";
9     pre=con.prepareStatement(sql);
10    pre.setInt(1, controlYear);
11    pre.setInt(2, areaId);
12    pre.setInt(3, categoryId);
13    rs=pre.executeQuery();
14  }
```

如果 categoryId 的值为 0 或 −1(即不限批次),则我们在 SQL 语句中添加一段排序的语句即可,然后使用 Connection 类型的 con 变量调用 prepareStatement(sql)方法来编译 SQL 语句。之后使用 PreparedStatement 类型的 pre 变量调用 setInt()方法设置变量的值。需要注意的是,setInt()方法的索引号要对应 SQL 语句中问号的位置。例如要设置 controlYear 的值,它在 SQL 语句中是第一个问号的位置,因此索引号就是 1。设置好值之后,我们就调用 pre 变量的 executeQuery()方法来执行 SQL 语句,并将结果存入结果集 rs 变量中。其他几个 if 嵌套也是一样的道理。

现在所有查询的结果都存放在结果集 rs 中,下一步我们需要把 rs 中的内容放入我们的 List 对象 controlLine 变量中。因此我们需要学习一个 while 循环,一条数据一条数据地保存,如程序清单 6-13 所示。

程序清单 6-13:Code0603\bktService\src\cn\jxufe\iet\impl\ControlLineImpl.java

```
1    while (rs.next()) {                              //判断是否还有记录
2        ControlLine line = new ControlLine();       //将一行记录转换成一个 Control 对象
3        line.setAreaName(rs.getString("areaName"));
4        line.setBatchName(rs.getString("batchName"));
5        line.setCategoryName(rs.getString("categoryName"));
6        line.setControlYear(rs.getInt("controlYear"));
7        line.setControlLine(rs.getInt("controlLine"));
8        controlLine.add(line);
9    }
```

这段代码中通过 rs 的 next() 方法来判断是否还有记录。每一次循环都新建一个 ControlLine 对象来保存记录。然后通过 rs 对象的 getString() 方法获取 ControlLine 对象的元素名,并调用 set 方法进行赋值。最后再把每一条记录通过 List 对象的 add() 方法添加进 List 链表中。因此,我们从数据库查询数据的过程还没有彻底完成,最后需要将获取的数据库连接关闭才行,使用 Connection 对象 con 调用 close() 方法即可。我们可以看到,代码中有很多 try{}catch{} 块,这是为了防止程序出现异常导致崩溃,是为了程序可以正常地运行下去。

ControlLineImpl 类的代码如程序清单 6-14 所示。

程序清单 6-14:Code0603\bktService\src\cn\jxufe\iet\impl\ControlLineImpl.java

```
1    package drhelp.cn.impl;
2    import java.sql.Connection;
3    import java.sql.PreparedStatement;
4    import java.sql.ResultSet;
5    import java.sql.SQLException;
6    import java.util.ArrayList;
7    import java.util.List;
8    import drhelp.cn.bean.ControlLine;
9    import drhelp.cn.util.DBconn;
10   /*
     从数据库获取省控线
11   */
12   public class ControlLineImpl {
13       public List<ControlLine> getControLine(int controlYear,int areaId,int categoryId,int batchId ){
14           Connection con=null;
15           ResultSet rs=null;
```

```
16          PreparedStatement pre=null;
17          DBconn db=new DBconn();
18          con=db.getConnection();
19          List<ControlLine> controlLine=new ArrayList<ControlLine>();
20          String sql="select controlYear,controlLine,areaName,batchName,
     categoryName from control_line l,area a,batch b,category c where l.areaId=
     a.areaId and l.batchId = b.batchId and l.categoryId = c.categoryId and
     controlYear=? and l.areaId=?";
21          if(batchId==0||batchId==-1){              //批次不限
22              if(categoryId==0||categoryId==-1){    //类别不限
23                  sql+=" order by l.categoryId";    //查询结果按照类别排序
24                  try {
25                      pre=con.prepareStatement(sql);
26                      pre.setInt(1, controlYear);
27                      pre.setInt(2, areaId);
28                      rs=pre.executeQuery();
29                  } catch (SQLException e) {
30                      printStackTrace();
31                  }
32              }else{
33                  sql+=" and l.categoryId=? order by l.categoryId";
34                  try {
35                      pre=con.prepareStatement(sql);
36                      pre.setInt(1, controlYear);
37                      pre.setInt(2, areaId);
38                      pre.setInt(3, categoryId);
39                      rs=pre.executeQuery();
40                  } catch (SQLException e) {
41          //TODO Auto-generated catch block
42                      printStackTrace();
43                  }
44              }
45          }else{
46              sql+=" and l.batchId=? order by l.categoryId";
47              try {
48                  pre=con.prepareStatement(sql);
49                  pre.setInt(1, controlYear);
50                  pre.setInt(2, areaId);
51                  pre.setInt(3, batchId);
52                  rs=pre.executeQuery();
53              } catch (SQLException e) {
54          //TODO Auto-generated catch block
55                  e.printStackTrace();
56              }
```

```
57        if(categoryId==0||categoryId==-1){        //类别不限
58        }else{
59            sql+=" and l.categoryId=? order by l.categoryId";
60        try {
61            pre=con.prepareStatement(sql);
62            pre.setInt(1, controlYear);
63            pre.setInt(2, areaId);
64            pre.setInt(3, batchId);
65            pre.setInt(4, categoryId);
66            rs=pre.executeQuery();
67        } catch (SQLException e) {
68            e.printStackTrace();
69        }
70  }
71  }//根据请求参数创建对应的SQL语句
72        try {
73          while (rs.next()) {                       //判断是否还有记录
74            ControlLine line = new ControlLine();   //将一行记录转换成一个
                                                      //Control对象
75            line.setAreaName(rs.getString("areaName"));
76            line.setBatchName(rs.getString("batchName"));
77            line.setCategoryName(rs.getString("categoryName"));
78            line.setControlYear(rs.getInt("controlYear"));
79            line.setControlLine(rs.getInt("controlLine"));
80            controlLine.add(line);
81        }
82        } catch (SQLException e) {
83  //TODO Auto-generated catch block
84            e.printStackTrace();
85        }finally{
86        try {
87            con.close();
88        } catch (SQLException e) {
89          e.printStackTrace();
90            }
91        }
92        return controlLine;
93        }
94  }
```

写完ControlLineImpl类后,我们下一步测试程序是否能够正确运行。在drhelp.cn.test包中建立测试类Test.java。Test类的代码如程序清单6-15所示。

程序清单6-15: Code0603\bktService\src\cn\jxufe\iet\test\Test.java

```
1   package drhelp.cn.test;
```

```
2    import java.util.ArrayList;
3    import java.util.List;
4    import drhelp.cn.bean.ControlLine;
5    import drhelp.cn.impl.ControlLineImpl;
6    public class Test {
7        public static void main(String[] args) {
8        List<ControlLine> controlLines=new ArrayList<ControlLine>();
9        ControlLineImpl cl=new ControlLineImpl();
10       controlLines=cl.getControLine(2011, 13, -1, -1);
11       for(ControlLine controlline:controlLines)
12       { System.out.println("省份:"+controlline.getAreaName()
13       +" 年份:"+controlline.getControlYear()
14       +" 批次:"+controlline.getBatchName()
15       +" 省控线:"+controlline.getControlLine()
16       +" 科类:"+controlline.getCategoryName());
17       }
18       }
19   }
```

在 Test 类中，我们直接调用 ControlLineImpl 类的 getControlLine 方法，并随便传入 4 个数，这里查询 2011 年江西的省控线（不限批次和类别）。传入的 controlYear 为 2011，areaId 为 13，categoryId 和 batchId 为 -1。然后再用 for 循环将结果打印出来。运行 Test.java 文件的结果如图 6-31 所示。

```
连接数据库成功!
省份:江西 年份:2011 批次:三本 省控线:413 科类:文史
省份:江西 年份:2011 批次:二本 省控线:484 科类:文史
省份:江西 年份:2011 批次:专科 省控线:260 科类:文史
省份:江西 年份:2011 批次:一本 省控线:532 科类:文史
省份:江西 年份:2011 批次:二本 省控线:474 科类:理工
省份:江西 年份:2011 批次:专科 省控线:200 科类:理工
省份:江西 年份:2011 批次:一本 省控线:531 科类:理工
省份:江西 年份:2011 批次:三本 省控线:343 科类:理工
```

图 6-31　测试结果图

6.6.3　练习

在本节中，我们只实现了如何在数据库中查询省控线信息。通过本节的学习，希望你能够自己接着完成"查询学校录取线"和"专业录取线"模块。

6.7　设计 App 服务器 Servlet 类

前面已经学习了如何从数据库中获取数据，但是 Android 客户端如何能够获取这些数据呢？这就要通过 Servlet 来实现。6.4 节详细介绍了一些主要的 Servlet 技术，还做了一个简单的小程序来加深理解。因此在本节中，我们更多的是接着之前的内容进行程序的编写。

6.7.1 省控线 Servlet 类的实现

Android 客户端是通过 JSON 格式来获取服务器数据。JSON（JavaScript Object Notation）是一种轻量级的数据交换格式，采用完全独立于语言的文本格式。JSON 表示形式为"名称/值对"：{"firstName":"value"}。这相当于 firstName＝Value。我们在前面建立实体类时都重写了 toString 方法，将数据以 JSON 格式输出。在服务器中，我们使用 Gson 来解析 JSON 数据。Gson 是 Google 提供的用来在 Java 对象和 JSON 数据之间进行映射的 Java 类库，它可以将一个 JSON 字符串转换成一个 Java 对象，也能够将一个 Java 对象转换成 JSON 字符串。要使用 Gson，就需要导入 Gson 类库。

我们在之前的项目上接着进行代码的编写。首先将 Gson 类库放入项目的 WebRoot\WEB-INF\lib 文件夹下，如图 6-32 所示。

在项目中建立名为 drhelp.cn.servlet 的包，用来专门存放所有的 Servlet 类。在包下建立 Servlet 类 ControlLineServlet，如图 6-33 所示。

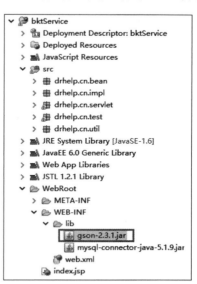

图 6-32　导入 Gson 类库

图 6-33　新建 ControlLineServlet 类

新建好 ControlLineServlet 后,我们开始编写 ControlLineServlet 类中的代码。

我们知道,doGet 方法和 doPost 方法分别用来处理客户端提出的 Get 请求和 Post 请求。但是 Android 客户端发送请求都是通过 Get 方法完成的,因此我们只需要在 doGet 方法中编写代码即可。如果你想在 doPost 方法中编写,也只需要在 doGet 方法中添加如下代码即可。

this.doPost(request, response);

它的意思是在 doGet 方法中调用 doPost 方法。添加了这条代码后,我们可直接在 doPost 方法中编写代码。我们先看 doPost 中的代码,如程序清单 6-16 所示。

程序清单 6-16:Code0604\bktService\src\drhelp\cn\servlet\ControlLineServlet.java

```
1   protected void doPost (HttpServletRequest request, HttpServletResponse
    response) throws ServletException, IOException {
2       response.setCharacterEncoding("UTF-8");        //设置返回结果的编码类型
3       PrintWriter out=response.getWriter();          //获取对应的输出流
4       ControlLineImpl cl=new ControlLineImpl();
5       int areaId=Integer.parseInt(request.getParameter("areaId"));
6       int year=Integer.parseInt(request.getParameter("year"));
7       int categoryId=getInt(request,"categoryId");
8       int batchId=getInt(request,"batchId");         //获取客户端传递过来的参数
9       List<ControlLine> controlLines=cl.getControoLine(year, areaId,
        categoryId,   batchId);
10      Gson gson=new Gson();                          //创建 Gson 对象
11      String result=gson.toJson(controlLines);       //将查询结果对象转换成 JSON
                                                       //格式字符串
12      out.write(result);                             //输出结果内容
13  }
```

首先,我们通过 response 对象的 setCharacterEncoding()方法设置返回到 Android 客户端的数据的编码类型为 UTF-8,防止返回的数据性乱码。然后,我们定义一个 PrintWriter 对象,用于将查询结果输出,PrintWriter 是一个非常实用的输出流;再定义一个 ControlLineImpl 对象,就是我们上一节所编写的类,用于查询数据库。我们通过 request 对象的 getParameter()方法来获取从 Android 客户端传递过来的省控线年份 year、地区编号 areaId、批次编号 batchId 和类别编号 categoryId。因为传递过来的数据都是 String 类型,所以我们需要用 Integer 对象的 parseInt()方法将数据强制转换为 Int 类型。前面说到,因为批次和类别的不确定因素,查询省控线一共有 4 种情况,所以我们单独编写 getInt 方法来处理批次和类别或许为空的情况。getInt 方法的代码如程序清单 6-17 所示。

程序清单 6-17:Code0604\bktService\src\drhelp\cn\servlet\ControlLineServlet.java

```
1   public static int getInt(HttpServletRequest request,String name){
    //根据请求参数获取对应的值
```

```
2       int result;
3       String nameString=request.getParameter(name);
4       if(nameString==null||"".equals(nameString)){//如果没有传递参数,默认为-1
5         result=-1;
6       }else{
7         try{
8           result=Integer.parseInt(nameString);
9         }catch(NumberFormatException ne){
10          result=-1;
11        }
12      }
13      return result;
14    }
```

在 getInt 方法中，我们在变量为空的情况下将它赋值为−1。

如此我们就得到了年份、地区编号、批次编号和类别编号 4 个变量，通过将这些变量传入我们所编写的 ControlLineImpl 对象的 getControLine() 方法中，就能够查询省控线。最后一步就是将我们的查询数据通过 Gson 转换为 JSON 数据输出。转换代码如程序清单 6-18 所示。

程序清单 6-18：Code0604\bktService\src\drhelp\cn\servlet\ControlLineServlet.java

```
1    Gson gson=new Gson();                              //创建 Gson 对象
2    String result=gson.toJson(controlLines);           //将查询结果对象转换成 JSON 格式字
                                                        //符串
3    out.write(result);                                 //输出结果内容
```

直接新建 Gson 对象，调用 toJson 方法进行转换，再使用 PrintWriter 对象的 write 方法输出结果。这样我们的 Android 客户端就能够接收到省控线信息。

ControlLineServlet 类的代码如程序清单 6-19 所示。

程序清单 6-19：Code0604\bktService\src\drhelp\cn\servlet\ControlLineServlet.java

```
1    package drhelp.cn.servlet;
2    import java.io.IOException;
3    import java.io.PrintWriter;
4    import java.util.List;
5    import javax.servlet.ServletException;
6    import javax.servlet.annotation.WebServlet;
7    import javax.servlet.http.HttpServlet;
8    import javax.servlet.http.HttpServletRequest;
9    import javax.servlet.http.HttpServletResponse;
10   import com.google.gson.Gson;
11   import drhelp.cn.bean.ControlLine;
12   import drhelp.cn.impl.ControlLineImpl;
```

```java
13  /**
14   * Servlet implementation class ControlLineServlet
15   */
16  @WebServlet("/ControlLineServlet")
17  public class ControlLineServlet extends HttpServlet {
18      private static final long serialVersionUID = 1L;
19      public ControlLineServlet() {
20          super();
21          //TODO Auto-generated constructor stub
22      }
23      protected void doGet(HttpServletRequest request, HttpServletResponse response) throws ServletException, IOException {
24          this.doPost(request, response);
25      }
26       protected void doPost (HttpServletRequest request, HttpServletResponse response) throws ServletException, IOException {
27          response.setCharacterEncoding("UTF-8");      //设置返回结果的编码类型
28          PrintWriter out=response.getWriter();        //获取对应的输出流
29          ControlLineImpl cl=new ControlLineImpl();
30          int areaId=Integer.parseInt(request.getParameter("areaId"));
31          int year=Integer.parseInt(request.getParameter("year"));
32          int categoryId=getInt(request,"categoryId");
33          int batchId=getInt(request,"batchId");       //获取客户端传递过来的参数
34          List<ControlLine> controlLines=cl.getControoLine(year, areaId, categoryId, batchId);
35          Gson gson=new Gson();                        //创建Gson对象
36          String result=gson.toJson(controlLines);
            //将查询结果对象转换成JSON格式字符串
37          out.write(result);                           //输出结果内容
38
39      }
40      public static int getInt(HttpServletRequest request,String name){
        //根据请求参数获取对应的值
41          int result;
42          String nameString=request.getParameter(name);
43          if(nameString==null||"".equals(nameString)){
        //如果没有传递参数,默认为-1
44              result=-1;
45          }else{
46              try{
47                  result=Integer.parseInt(nameString);
48              }catch(NumberFormatException ne){
49                  result=-1;
50              }
```

```
51          }
52          return result;
53      }
54  }
```

接下来我们通过测试查看是否成功,将项目部署到 Tomcat 中并运行。

① 指定批次、指定科类。查询地区编号为 13、科类编号为 1、批次编号为 1、年份为 2011 年的省控线。

打开浏览器,输入:

```
http://localhost:8080/bktService/ControlLineServlet?areaId=13&categoryId=1&batchId=1&year=2011
```

出现图 6-34 所示的结果。

图 6-34　限批次、限科类测试结果

② 批次不限、指定科类。查询地区编号为 13、科类编号为 1、年份为 2011 年的省控线。

打开浏览器,输入:

```
http://localhost:8080/bktService/ControlLineServlet?areaId=13&categoryId=1&year=2011
```

出现图 6-35 所示的结果。

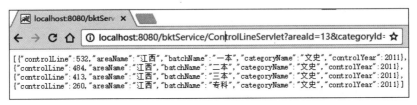

图 6-35　限科类测试结果

③ 不限批次、不限科类。查询地区编号为 13,年份为 2011 年的省控线。

打开浏览器,输入:

```
http://localhost:8080/bktService/ControlLineServlet?areaId=13&categoryId=1&year=2011
```

出现图 6-36 所示的结果。

6.7.2　练习

上一节从数据库中查询学校录取线和专业录取线的模块做完了吗?本节需要你继续

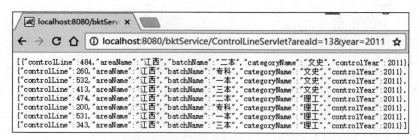

图 6-36　不限批次、不限科类测试结果

完成这两个模块的 Servlet 类的编写。

6.8　App 服务器端设计巩固

通过本章前面章节的学习,你已经掌握了服务器开发的主要知识点,也已经具备了服务器开发的基本能力。本节单独将"豹考通"问题模块拿出来,是为了让你对现有知识进行一个全面的巩固和提升。

6.8.1　问题模块实体类

问题实体类需要有问题标记、问题内容/描述、提问时间、问题回复时间和问题状态(暂未回复、未审核、已通过)。因此,可以确定问题实体类的成员变量,各字段含义如下。

- questionId:问题标记。
- questionContent:问题内容/描述。
- questionTime:提问时间。
- answerContent:问题回复内容。
- answerTime:问题回复时间。
- questionStatus:问题状态(暂未回复、未审核、已通过)。

在上一节项目的基础之上,在 drhelp.cn.bean 包中建立问题实体类 Question,并且重写 Question 的 toString 方法(见程序清单 6-20)。

程序清单 6-20：Code060501\bktService\src\cn\jxufe\iet\bean\Question.java

```
1   package drhelp.cn.bean;
2   import java.io.Serializable;
3   public class Question implements Serializable {    //报考咨询中的问题
4       private static final long serialVersionUID = 1L;
5       private int questionId;                         //问题 id
6       private String questionContent;                 //问题内容
7       private String questionTime;                    //提问时间
8       private String answerContent;                   //问题回复
9       private String answerTime;                      //回复时间
10      private String questionStatus;                  //问题状态:待回答、已回答
```

```java
11  public int getQuestionId() {
12      return questionId;
13  }
14  public void setQuestionId(int questionId) {
15      his.questionId = questionId;
16  }
17  public String getQuestionContent() {
18      return questionContent;
19  }
20  public void setQuestionContent(String questionContent) {
21      this.questionContent = questionContent;
22  }
23  public String getQuestionTime() {
24      return questionTime;
25  }
26  public void setQuestionTime (String questionTime) {this. questionTime = questionTime;
27  }
28  public String getAnswerContent() {
29      return answerContent;
30  }
31  public void setAnswerContent(String answerContent) {
32      this.answerContent = answerContent;
33  }
34  public String getAnswerTime() {
35      return answerTime;
36  }
37  public void setAnswerTime(String answerTime) {
38      this.answerTime = answerTime;
39  }
40  public String getQuestionStatus() {
41      return questionStatus;
42  }
44  public void setQuestionStatus(String questionStatus) {
44      this.questionStatus = questionStatus;
45  }
46  @Override
47  public String toString() {
48    return "Question [questionId=" + questionId +
49  ", questionContent="+ questionContent +
50  ", questionTime=" + questionTime+
51  ", answerContent=" + answerContent +
52  ", answerTime="+ answerTime +
53  ", questionStatus=" + questionStatus + "]";
```

```
54    }
55  }
```

6.8.2 问题模块数据库操作类

"豹考通"问题模块包括的功能有发布问题、回复问题和查询问题,因此涉及的数据操作有数据库查询、插入和更新。

在 drhelp.cn.impl 包下建立问题模块的数据库操作类 QuestionImpl。

1. 查询问题

服务器根据从 Android 客户端传递过来的关键字查询相关问题,支持模糊查询,如果没有关键字,则表示查询所有的问题。查询问题传递的参数有关键字(keyword)。查询问题模块的代码如程序清单 6-21 所示。

程序清单 6-21：Code060501\bktService\src\cn\jxufe\iet\impl\QuestionImpl.java

```
1   public List<Question> getQuestions(String keyword){
2       List<Question> questions=new ArrayList<Question>();
3       Connection con=null;
4       ResultSet rs=null;
5       PreparedStatement pre=null;
6       DBconn db=new DBconn();
7       con=db.getConnection();
8       String sql;
9       if(keyword==null||"".equals(keyword)){       //如果关键字为空
10          sql="select * from question";
11      try {
12          pre=con.prepareStatement(sql);
13          rs=pre.executeQuery();
14      } catch (SQLException e) {
15          printStackTrace();
16      }
17      }else{
18          sql="select * from question where questionContent like ?";
19      try {
20          pre=con.prepareStatement(sql);
21          pre.setString(1, "%"+keyword+"%");
22          rs=pre.executeQuery();
23      } catch (SQLException e) {
24          printStackTrace();
25      }
26      }
27      try {
28          while (rs.next()) {                      //判断是否还有记录
29              Question question=new Question();
```

```
30        question.setQuestionId(rs.getInt("questionId"));
31        question.setQuestionContent(rs.getString("questionContent"));
32        question.setQuestionStatus(rs.getString("questionStatus"));
33        question.setQuestionTime(rs.getString("questionTime"));
34        question.setAnswerContent(rs.getString("answerContent"));
35        question.setAnswerTime(rs.getString("answerTime"));
36        questions.add(question);
37      }
38    } catch (SQLException e) {
39        e.printStackTrace();
40    }finally{
41    try {
42        con.close();
43    } catch (SQLException e) {
44        e.printStackTrace();}
45    }
46    return questions;
47  }
```

在查询问题的代码中，主要是用 if 语句来动态编辑 SQL 查询语句。对传入的关键字 keyword 进行判断，如果为空，则 sql="select * from question"将数据库中所有问题查询出来，然后再执行 SQL 语句，并把结果存入结果集 rs 变量中；如果不为空，则是 sql="select * from question where questionContent like ?"。需要注意的是，我们使用 like 关键字来进行模糊查询。因此在设置 SQL 语句的变量值的使用需要在关键字前后添加％符号，例如代码中的 pre.setString(1,"％"＋keyword＋"％")，否则无法进行模糊查询。在查询完成后，我们再使用 while 循环将查询得到的结果集 rs 存入 Question 实体对象中。注意，开启数据库连接后，需要在使用后关闭连接。

同样地，我们写好的查询方法也在 Test.java 中测试一下。测试代码如程序清单 6-22 所示。

程序清单 6-22：Code060501\bktService\src\cn\jxufe\iet\test\Test.java

```
1   List<Question> questions=new ArrayList<Question>();
2   QuestionImpl qi=new QuestionImpl();
3   questions=qi.getQuestions(null);
4   for(Question question: questions){System.out.println("问题内容:"+
    question.getQuestionContent());
5   }
```

我们将 null 传入查询方法中，使用 for 循环打印每条问题的内容。运行 Test.java 文件后得到图 6-37 所示的结果。

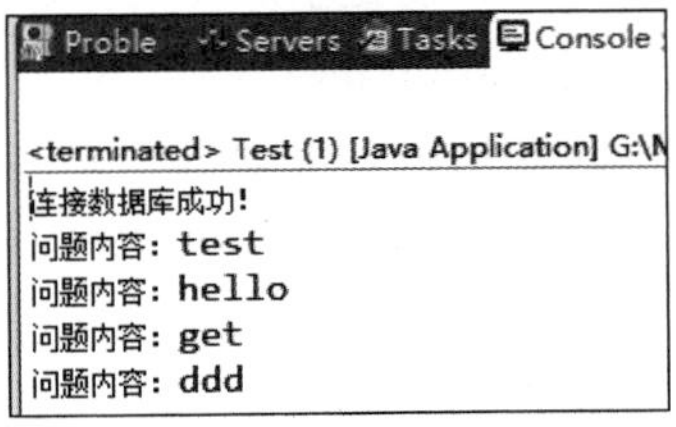

图 6-37　测试查询问题方法

2. 提交问题

Android 客户端将封装好问题信息,以 JSON 格式字符串进行传递,服务器端将 JSON 格式字符串转换成问题对象,然后插入数据库中,服务器只需要返回一个是否成功插入的标识给客户端即可,因此我们所建立的方法是 boolean 类型。需要传递的数据有用于问题信息封装的 JSON 字符串(question),提交的问题信息包括问题内容和提问时间。提交问题模块的代码如程序清单 6-23 所示。

程序清单 6-23:Code060501\bktService\src\cn\jxufe\iet\impl\QuestionImpl.java

```java
1   public boolean insertQuestion(String questionString){
2       boolean flag = false;
3       Gson gson=new Gson();
4       Question question=gson.fromJson(questionString, Question.class);
5       Connection con=null;
6       PreparedStatement pre=null;
7       DBconn db= new DBconn();
8       con=db.getConnection();
9       String sql="insert into question(questionContent,questionTime,
    questionStatus)values(?,?,?)";
10      try {
11          pre= con.prepareStatement(sql);
12          pre.setString(1, question.getQuestionContent());
13          pre.setString(2, question.getQuestionTime());
14          pre.setString(3, "暂未回复");
15          int i = pre.executeUpdate();              //执行更新操作
16          if (i == 1) {                             //如果成功,则影响1行
17              flag = true;
18          }
19          con.close();
20      } catch (SQLException e) {
21          printStackTrace();
22      }
23      return flag;
24  }
```

首先定义 boolean 类型变量 flag,并初始化为 false,因为 Android 客户端传递过来的 questionString 是 JSON 类型数据,所以我们使用 Gson 的 fromJson 方法将 questionString 转换为 Question 的 Java 对象类型。然后通过 Question 对象的 get 方法获取问题内容和提问时间并赋值给 SQL 语句。调用 PreparedStatement 对象的 executeUpdate()方法执行更新。通过 executeUpdate()返回的标识来进行判断;若为 1,则将 flag 赋值为 ture,说明更新成功。

3. 回复问题

Android 客户端将封装好问题信息,以 Gson 格式字符串进行传递,服务器端将 Gson

格式字符串转换成问题对象，然后更新数据库中相关的记录。与提交问题一样，服务器只需要回馈给 Android 客户端是否成功回复问题的一个标识即可，因此回复问题的方法也是 boolean 类型。需要传递的数据有用于问题信息封装的 Gson 字符串（question），回复问题信息包括问题 id、问题内容、提问时间、回复内容、回复时间等。回复问题模块的代码如程序清单 6-24 所示。

程序清单 6-24：Code060501\bktService\src\cn\jxufe\iet\impl\QuestionImpl.java

```java
1    public boolean updateQuestion(String questionString){
2        boolean flag = false;
3        Gson gson=new Gson();
4        Question question=gson.fromJson(questionString, Question.class);
5        Connection con=null;
6        PreparedStatement pre=null;
7        DBconn db=new DBconn();
8        con=db.getConnection();
9        String sql="update question set answerContent=?, answerTime=?, questionStatus=? where questionId=?";
10       try {
11           pre= con.prepareStatement(sql);
12           pre.setString(1, question.getAnswerContent());
13           pre.setString(2, question.getAnswerTime());
14           pre.setString(3, "未审核");
15           pre.setInt(4, question.getQuestionId());
16           int i = pre.executeUpdate();              //执行更新操作
17           if (i == 1) {                             //如果成功，则影响 1 行
18               flag = true;
19           }
20           con.close();
21       } catch (SQLException e) {
22           e.printStackTrace();
23       }
24       return flag;
25   }
```

此模块和提交问题模块思路大致相通，都是通过 Gson 将 JSON 数据转换为 Question 的 Java 对象再进行操作，这里不再细说。

由此，问题模块的数据库操作类已经编写完成。

QuestionImpl 类的代码如程序清单 6-25 所示。

程序清单 6-25：Code060501\bktService\src\cn\jxufe\iet\impl\QuestionImpl.java

```java
1    package drhelp.cn.impl;
2    import java.sql.Connection;
```

```
3    import java.sql.PreparedStatement;
4    import java.sql.ResultSet;
5    import java.sql.SQLException;
6    import java.util.ArrayList;
7    import java.util.List;
8    import com.google.gson.Gson;
9    import drhelp.cn.bean.Question;
10   import drhelp.cn.util.DBconn;
11   public class QuestionImpl {
12   /*
     查询问题
13   */
14       public List<Question> getQuestions(String keyword){
15       List<Question> questions=new ArrayList<Question>();
16       Connection con=null;
17       ResultSet rs=null;
18       PreparedStatement pre=null;
19       DBconn db=new DBconn();
20       con=db.getConnection();
21       String sql;
22       if(keyword==null||"".equals(keyword)){    //如果关键字为空
23         sql="select * from question";
24       try {
25         pre=con.prepareStatement(sql);
26         rs=pre.executeQuery();
27       } catch (SQLException e) {
28         printStackTrace();
29       }
30       }else{
31         sql="select * from question where questionContent like ?";
32       try {
33         pre=con.prepareStatement(sql);
34         pre.setString(1, "%"+keyword+"%");
35         rs=pre.executeQuery();
36       } catch (SQLException e) {
37         printStackTrace();
38       }
39       }
40       try {
41         while (rs.next()) {                     //判断是否还有记录
42           Question question=new Question();
43           question.setQuestionId(rs.getInt("questionId"));
44           question.setQuestionContent(rs.getString("questionContent"));
45           question.setQuestionStatus(rs.getString("questionStatus"));
```

```
46              question.setQuestionTime(rs.getString("questionTime"));
47              question.setAnswerContent(rs.getString("answerContent"));
48              question.setAnswerTime(rs.getString("answerTime"));
49              questions.add(question);
50          }
51       } catch (SQLException e) {
52         e.printStackTrace();
53       }finally{
54        try {
55         con.close();
56       } catch (SQLException e) {
57          e.printStackTrace();
58       }
59       }
60       return questions;
61    }
62    /*
       插入问题
63     */
64     public boolean insertQuestion(String questionString){
65        boolean flag = false;
66        Gson gson=new Gson();
67        Question question=gson.fromJson(questionString, Question.class);
68        Connection con=null;
69        PreparedStatement pre=null;
70        DBconn db=new DBconn();
71        con=db.getConnection();
72        String sql="insert into Question
73          (questionContent,questionTime,questionStatus)values(?,?,?)";
74     try {
75        pre=con.prepareStatement(sql);
76        pre.setString(1, question.getQuestionContent());
77        pre.setString(2, question.getQuestionTime());
78        pre.setString(3, "暂未回复");
79        int i = pre.executeUpdate();              //执行更新操作
80     if (i == 1) {                                //如果成功,则影响1行,否则
81        flag = true;
82     }
83     con.close();
84     } catch (SQLException e) {
85        e.printStackTrace();
86  }
87  return flag;
88  }
```

```
89    /*
      更新问题
90    */
91    public boolean updateQuestion(String questionString){
92        boolean flag = false;
93        Gson gson=new Gson();
94        Question question=gson.fromJson(questionString, Question.class);
95        Connection con=null;
96        PreparedStatement pre=null;
97        DBconn db=new DBconn();
98        con=db.getConnection();
99        String sql="update question set answerContent=?, answerTime=?, questionStatus=? where questionId=?";
100       try {
101         pre=con.prepareStatement(sql);
102         pre.setString(1, question.getAnswerContent());
103         pre.setString(2, question.getAnswerTime());
104         pre.setString(3, "未审核");
105         pre.setInt(4, question.getQuestionId());
106         int i = pre.executeUpdate();              //执行更新操作
107         if (i == 1) {                             //如果成功,则影响1行,否则
108           flag = true;
109         }
110         con.close();
111       } catch (SQLException e) {
112         e.printStackTrace();
113       }
114       return flag;
115   }
116   }
```

6.8.3 问题模块 Servlet 类

在问题模块的 Servlet 类中,我们都知道只有一个 doPost 或 doGet 方法,可是我们的问题模块具有 3 个功能,分别是提交问题、回复问题和查询问题。那么如何能够每一次都按需求自己选择执行对应的功能呢?其实要解决这个问题不难,我们只需要让 Android 客户端也传递一个标识变量,然后在 doPost 或 doGet 方法中使用 if 语句判断这个标识变量,之后运行相对应的程序代码即可。

首先在 drhelp.cn.servlet 包中新建一个问题模块的 Servlet 类 QuestionServlet。
我们将代码写在 doPost 方法中,那么 doGet 方法中就需要添加如下代码:

this.doPost(request, response)

doPost 中的代码如程序清单 6-26 所示。

程序清单 6-26：Code060502\bktService\src\cn\jxufe\iet\servlet\QuestionServlet.java

```
1   protected void doPost (HttpServletRequest request, HttpServletResponse
    response) throws ServletException, IOException {
2       response.setCharacterEncoding("UTF-8");
3       PrintWriter out = response.getWriter();
4       String flag=request.getParameter("flag");
5       QuestionImpl qi=new QuestionImpl();
6       if("insert".equals(flag)){                    //提交问题
7           Boolean result=qi.insertQuestion(request.getParameter("question"));
8           out.write(result+"");
9       }else if("update".equals(flag)){              //回复问题
10          boolean result=qi.updateQuestion(request.getParameter("question"));
11          out.write(result+"");
12      }else{                                        //查询问题
13          List<Question> questions=qi.getQuestions(request
    .getParameter("keyword"));
14          Gson gson=new Gson();
15          String result=gson.toJson(questions);
16          out.write(result);
17      }
18  }
```

首先定义一个 String 类型 flag 来存放 Android 客户端传递过来的标识变量（update 为回复问题；insert 为提交问题；其他默认为查询问题），我们使用 if 语句对 flag 进行判断，并在 if 中调用不同的方法进行不同的操作。

至此，问题模块的 Servlet 类也编写好了。接下来，我们将项目部署到 Tomcat 中进行测试。

打开浏览器，输入 http://localhost:8080/bktService/QuestionServlet，我们不传入 flag 标识变量是代表查询所有问题。结果如图 6-38 所示。

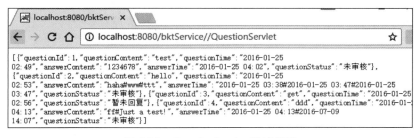

图 6-38 QuestionServlet 测试结果

QuestionServlet 类的代码如程序清单 6-27 所示。

程序清单6-27：Code060502\bktService\src\cn\jxufe\iet\Servlet\QuestionServlet.java

```java
1   package drhelp.cn.servlet;
2   import java.io.IOException;
3   import java.io.PrintWriter;
4   import java.util.List;
5   import javax.servlet.ServletException;
6   import javax.servlet.annotation.WebServlet;
7   import javax.servlet.http.HttpServlet;
8   import javax.servlet.http.HttpServletRequest;
9   import javax.servlet.http.HttpServletResponse;
10  import com.google.gson.Gson;
11  import drhelp.cn.bean.Question;
12  import drhelp.cn.impl.QuestionImpl;
13  /**
14   * Servlet implementation class QuestionServlet
15   */
16  @WebServlet("/QuestionServlet")
17  public class QuestionServlet extends HttpServlet {
18      private static final long serialVersionUID = 1L;
19  /**
20   * @see HttpServlet#HttpServlet()
21   */
22  public QuestionServlet() {
23      super();
24      //TODO Auto-generated constructor stub
25  }
26  /**
27   * @see HttpServlet#doGet(HttpServletRequest request, HttpServletResponse response)
28   */
29  protected void doGet (HttpServletRequest request, HttpServletResponse response) throws ServletException, IOException {
30      this.doPost(request, response);
31  }
32  /**
33   * @see HttpServlet#doPost(HttpServletRequest request, HttpServletResponse response)
34   */
35  protected void doPost (HttpServletRequest request, HttpServletResponse response) throws ServletException, IOException {
36      response.setCharacterEncoding("UTF-8");
37      PrintWriter out = response.getWriter();
38      String flag=request.getParameter("flag");
```

```
39        QuestionImpl qi=new QuestionImpl();
40        if("insert".equals(flag)){            //提交问题
41            boolean result=qi.insertQuestion(request.getParameter("question"));
42            out.write(result+"");
44        }else if("update".equals(flag)){       //更新问题
44            boolean result=qi.updateQuestion(request.getParameter("question"));
45            out.write(result+"");
46        }else{                                 //查询问题
47            List<Question> questions=qi.getQuestions(request
   .getParameter("keyword"));
48            Gson gson=new Gson();
49            String result=gson.toJson(questions);
50            out.write(result);
51        }
52    }
53 }
```

6.9 本章小结

本章围绕"豹考通"服务器端搭建讲解相关知识。服务器端搭建有多种技术，如 JSP、Spring 框架、Hibernate 框架等，本书中使用 JSP 和 Spring 框架。本章使用 JSP 实现，而 Spring 框架将在第 8 章中重点讲解。本章重点介绍了 Servlet、JavaBean、Tomcat、客户端与服务器端交互理论知识等。

在进行服务器端省控线查询、问题模块编程时，使用 JDBC 技术实现连接数据库，介绍了 Tomcat 服务器的安装及其使用、Servlet 的相关知识以及如何部署已开发好的 Java Web 应用程序。

6.10 课后练习

1. 完成 6.4.6 节的实战演练项目。
2. 参照 6.6 节，在数据库中查询省控线信息的模块设计，自行设计查询学校录取线和专业录取线的业务逻辑类。

第 7 章

App 客户端与服务器交互设计

7.1 本章简介

前面章节中已经实现了本地版"豹考通"的客户端和服务器端,客户端还没有和服务器端进行交互。本章主要实现客户端替换临时数据与服务器端进行交互,其中包括客户端和服务器端交互理论、"省控线查询"模块与服务器端交互实现过程、"历年录取线查询"模块与服务器端交互实现过程和"报考咨询"模块与服务器端交互实现过程。

本章内容在本地版客户端和网络服务器基础上完成。首先,在前面开发的客户端基础上把第三方 JAR 包导入;其次,在 util 文件夹下创建一些与服务器端交互用到的工具类;最后,分别实现与服务器端交互的"省控线查询""历年录取线查询"和"报考咨询"3 个模块,具体流程参见图 7-1。

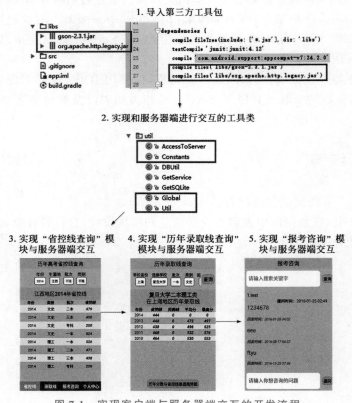

图 7-1 实现客户端与服务器端交互的开发流程

7.2 客户端和服务器端数据交互基础

7.2.1 HttpClient

Android 客户端与服务器端交互的方式通常有两种：一种是直接采用 Android 内置的 HttpClient 发送 HTTP 请求和获取 HTTP 响应；另一种是利用 ksoap2-android 项目，调用服务器端提供的 Web Service。"豹考通"项目中采用第一种方式，通过 HttpClient 与服务器端进行交互。

HttpClient 是 Apache 开源组织提供的一个项目，HttpClient 开发过程中涉及的类主要包含以下几个。

- **HttpClient**：HTTP 客户端接口，该接口封装了执行 HTTP 请求所需的各种对象，这些对象可以处理 Cookie、授权、连接管理以及其他一些特性。HttpClient 的线程安全依赖于具体的客户端的实现和配置。该接口中包含多种 execute() 方法，用于执行具体的请求，通常通过 DefaultHttpClient 子类来创建该接口的对象。
- **HttpGet**：该类采用 GET 方式发送 HTTP 请求，通常将请求的 URL 作为参数传给该类的构造函数。
- **HttpPost**：该类采用 POST 方式发送 HTTP 请求，通常将请求的 URL 作为参数传给该类的构造函数。
- **HttpResponse**：该接口封装了 HTTP 的相应信息，通过调用相应方法可以获取 HTTP 响应信息，例如获取响应状态、响应内容等。
- **HttpEntity**：HTTP 信息封装类，通过该类可以获取 HTTP 请求或响应的内容、长度、类型、编码方式等信息。HttpEntity 对象可以通过 HTTP 消息发送和接收，既可以存在于请求消息中，也可以存在于响应消息中。

这些类是如何协同工作来完成 HTTP 请求和获取响应信息的呢？通常我们会采用 GET 和 POST 两种方式来发送请求，而 GET 和 POST 发送请求各有优缺点，须根据具体的情境选择。两者的操作模式有所差别，主要是参数传递方式有所不同。因此，在通过 HttpClient 与服务器端交互时，需要针对这两种情况单独处理。使用 HttpClient 发送请求获取响应信息的流程如图 7-2 所示。

通过图中所述执行过程，我们可以总结出，无论是使用 HttpGet，还是使用 HttpPost，都必须通过以下 3 步来访问 HTTP 资源。

（1）创建 HttpGet 或 HttpPost 对象，将要请求的 URL 通过构造函数传入 HttpGet 或 HttpPost 对象。

（2）使用 DefaultHttpClient 类的 execute 方法发送 HTTP GET 或 HTTP POST 请求，并返回 HttpResponse 对象。

（3）通过 HttpResponse 接口的 getEntity 方法返回响应信息，并进行相应的处理。

注意：如果使用 HttpPost 方法提交 HTTP POST 请求，还需要使用 HttpPost 类的

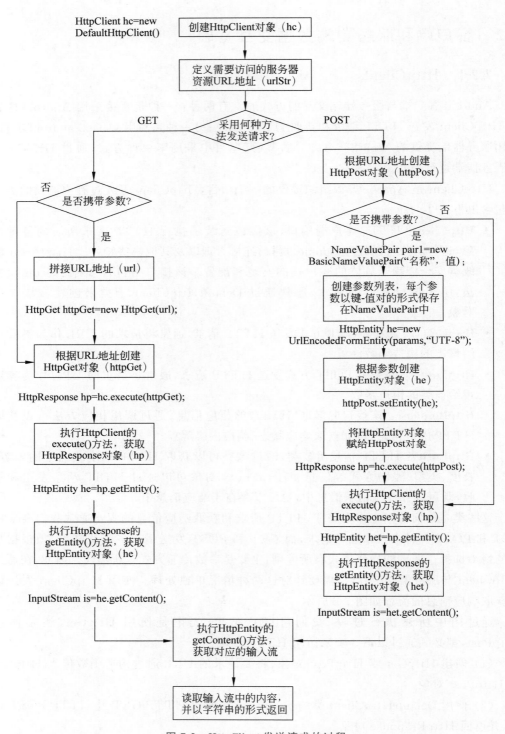

图 7-2　HttpClient 发送请求的过程

setEntity 方法设置请求参数。访问网络时,需要在清单文件中添加访问网络的权限。此外,在 Android 高版本(Android 3.0 以上)中,是将访问网络看成一种较为耗时的操作,不能直接在主线程中访问网络,需要单独启动线程或通过异步任务来访问网络。

7.2.2 JSON 解析

服务器端响应客户端请求返回的结果的表示方式有多种,可以是具体的 HTML 页面、存储结构化数据的 XML 文件、JSON 格式的字符串等。在"豹考通"项目中,客户端与服务器端数据交互采用 JSON 格式,通过谷歌提供的第三方 JAR 包(Gson)来实现 JSON 格式字符串的生成和解析。

JSON 是一种轻量级的数据交换格式,采用完全独立于语言的文本格式,既易于人阅读和编写,也易于机器解析和生成。JSON 中主要有两种数据结构。

- 由键-值对组成的数据结构,这种结构在不同语言中有不同的实现,例如,在 C 语言中是一个结构体(struct),在 Java 中是一种映射结构,在 JavaScript 中则是一个对象等。
- 有序集合,这种数据结构在不同语言中可能是列表、集合、数组等实现。

Java 对象与 JSON 数据之间的转换主要包含两种情况:一种是将 Java 对象转换成 JSON 格式,只需要调用 Gson 对象的 toJson()方法并将 Java 对象作为参数传递即可;另一种是将 JSON 格式的字符串转换成一个 Java 对象,也非常简单,只需要调用 Gson 对象的 fromJson()方法,传递两个参数(一个是需要转换的 JSON 格式的字符串,另一个是需要转换的对象的类型)。详细代码会在后面的具体模块实现部分列出。

7.2.3 第三方 JAR 包导入

"豹考通"客户端项目中主要使用了两个第三方 JAR 包:gson-2.3.1.jar 和 org.apache.http.legacy.JAR 包。导入第三方 JAR 包有两种方式,具体如下。

1. 方式一

(1) 下载需要的 JAR 包。

(2) 将 JAR 包复制到 Project 下的 app→libs 目录(没有 libs 目录就新建一个),如图 7-3 所示。

(3) 右击该 JAR 包,选择 Add As Library 选项,添加依赖包,弹出图 7-4 所示窗口,单击 OK 按钮即可。

(4) 打开 build.gradle 文件,在 dependencies 方法中出现 JAR 包名称,则说明导入成功,如图 7-5 所示。

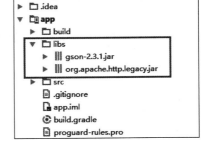

图 7-3 复制 JAR 包到 libs 文件夹

2. 方式二

(1) 下载需要的 JAR 包。

(2) 将 JAR 包复制到 Project 下的 app→libs 目录(没有 libs 目录就新建一个),如图 7-6 所示。

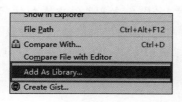 

图 7-4　添加依赖包

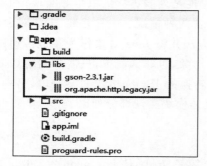

图 7-5　JAR 包导入成功

（3）单击工具栏中的 Project Structure 按钮，如图 7-7 所示。

图 7-6　复制 JAR 包到 libs 文件夹　　　图 7-7　单击 Project Structure 按钮

（4）选择 Dependencies 选项卡，单击右边的加号，选择 Jar Dependency 选项，如图 7-8 所示。

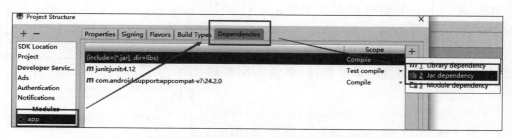

图 7-8　选择 Jar Dependency 选项

（5）在弹出的窗口中选择之前复制的 JAR 包，如图 7-9 所示。
（6）单击 OK 按钮即可，变成图 7-10 所示，就表明导入成功。

第 7 章　App 客户端与服务器交互设计

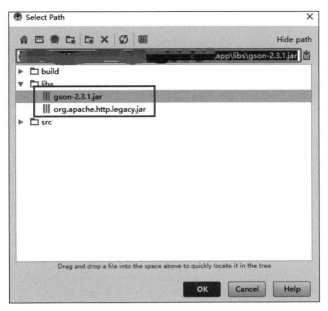

图 7-9　选择 JAR 包路径

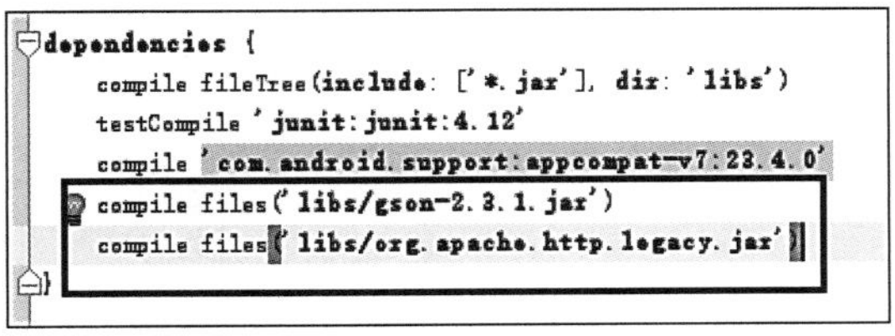

图 7-10　JAR 包导入成功

7.2.4　客户端与服务器端交互工具类设计

在实现每个模块与服务器端进行交互前,在文件夹 util 下创建一些工具类,包括 AccessToServer 类、Constants 类、Global 类和 Util 类,如图 7-11 所示。

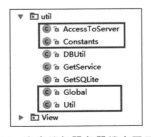

图 7-11　客户端与服务器端交互工具类

AccessToServer 类中含有 doGet 和 doPost 两个方法,调用这两个方法时,传入参数向服务器端发送请求,就能得到响应结果,代码如程序清单 7-1 所示。

程序清单 7-1：Code0701\app\src\main\java\cn\jxufe\iet\code0701\util\AccessToServer.java

```java
1   public class AccessToServer {
2       /**
3        * 向服务器发送 Get 请求,获取响应结果
4        * @param url                    表示需要访问的资源网址；
5        * @param names                  表示需要传递的多个参数名称集合；
6        * @param values                 表示传递的每个参数所对应的值；
7        * @return                       返回字符串(通常是 JSON 格式)
8        */
9       public String doGet(String url, String[] names, String[] values) {
10          String result = "";                     //保存返回结果
11          if (names != null) {
12              //当有参数时,将参数拼接在地址后面,并用?隔开,参数间用 & 隔开
13              url += "?";                          //在网址后面添加?符号
14              for (int i = 0; i < names.length; i++) {
15                  //循环遍历参数名和参数值,将其拼接
16                  url += names[i] + "=" + values[i];
17                  if (i != (names.length - 1)) {
18                      //如果不是最后一个参数,则添加 & 符号
19                      url += "&";
20                  }
21              }
22          }
23          System.out.println("url="+url);
24          HttpClient httpClient = new DefaultHttpClient();
                                                    //创建 HttpClient 对象
25          HttpGet httpGet = new HttpGet(url);     //创建 HttpGet 对象
26          try {
27              HttpResponse httpResponse = httpClient.execute(httpGet);
28              //发送请求获取响应
29              if (httpResponse.getStatusLine().getStatusCode() == HttpStatus.SC_OK) {
30                  //判断返回结果状态
31                  HttpEntity httpEntity = httpResponse.getEntity();
32                  //获取返回的实体
33                  InputStream is = httpEntity.getContent();
34                  //获取实体的内容
35                  result=readFromInputStream(is);   //读取输入流中的内容
36              } else {
37                  result = Constants.ERROR;         //返回出错
38              }
39          } catch (Exception e) {                   //访问过程中抛出异常
40              result=Constants.EXCEPTION;           //返回异常
```

```
41            }
42            return result;                              //返回结果
43       }
44       /**
45        * 向服务器发送 Post 请求,获取响应结果
46        * @param url    表示需要访问的资源网址;
47        * @param names  表示需要传递的多个参数名称集合;
48        * @param values 表示传递的每个参数所对应的值;
49        * @return       返回字符串(通常是 JSON 格式)
50        */
51       public String doPost(String url, String[] names, String[] values) {
52            String result = "";                         //保存返回结果
53            HttpClient httpClient = new DefaultHttpClient();
                                                          //创建 HttpClient 对象
54            HttpPost httpPost = new HttpPost(url);      //创建 HttpPost 对象
55            if (names != null) {                        //将传递的参数写入请求体中
56                ArrayList<NameValuePair> params = new ArrayList
   <NameValuePair>();                                    //创建键-值对集合
57                for (int i = 0; i < names.length; i++) {
58                    //循环遍历参数名和参数值,将其关联起来并添加到集合中去
59                    NameValuePair pair = new BasicNameValuePair(names[i],
   values[i]);
60                    //将关键字与对应的值关联起来
61                    params.add(pair);                   //将参数添加到集合中去
62                }
63                try {
64                    HttpEntity requestEntity = new UrlEncodedFormEntity(params,
65   "UTF-8");                                            //创建请求实体,传递参数
66                    httpPost.setEntity(requestEntity);  //将请求实体放入 Post 请
                                                          //求中去
67                } catch (Exception ex) {                //捕获异常
68                    result = Constants.EXCEPTION;       //返回异常
69                }
70            }
71            try {
72                HttpResponse httpResponse = httpClient.execute(httpPost);
                                                          //发送 Post 请求
73                if (httpResponse.getStatusLine().getStatusCode() ==
   HttpStatus.SC_OK) {
74                    //判断返回结果状态
75                    HttpEntity httpEntity = httpResponse.getEntity();
                                                          //获取返回的实体
76                    InputStream is = httpEntity.getContent();
                                                          //获取实体的内容
```

```
77              result=readFromInputStream(is);           //读取输入流中的内容
78          } else {
79              result=Constants.ERROR;
80          }
81      } catch (Exception e) {                           //捕获异常
82          result = Constants.EXCEPTION;                 //返回异常
83      }
84      return result;
85  }
86  private String readFromInputStream(InputStream is){
87      //读取输入流中的内容
88      String result="";                                 //保存读取的内容
89      byte[] buffer = new byte[1024];                   //定义缓存数组
90      int hasRead = 0;                                  //记录读取的字节数
91      try {
92          while ((hasRead = is.read(buffer)) != -1) {
93              //循环读取输入流中的内容,直到结束
94              result += new String(buffer, 0, hasRead);
95              //将读取的内容拼接成字符串
96          }
97      } catch (IOException e) {                         //读取过程中抛出异常
98          result = Constants.EXCEPTION;                 //返回异常
99      }
100     return result;
101 }
102 }
```

在 Constants 工具类中定义一些常量,主要是在"豹考通"模块中启动线程中发送的常量信息,具体代码如程序清单 7-2 所示。

程序清单 7-2:Code0701\app\src\main\java\cn\jxufe\iet\code0701\util\Constants.java

```
1   public class Constants {                                        //定义一些常量
2     public static final int START_ANIMATION=0x00;                 //开启动画
3     public static final int STOP_ANIMATION=0x01;                  //停止动画
4     public static final int GET_SCHOOL_ENROLL_SCORE=0x11;         //获取学校录取线
5     public static final int GET_MAJOR_ENROLL_SCORE=0x12;          //获取专业录取线
6     public static final int GET_CONTROL_LINE=0x13;                //获取省控线
7     public static final int GET_QUESTIONS=0x14;                   //获取问题
8     public static final int SUBMIT_QUESTION=0x15;                 //提交问题
9     public static final int UPDATE_QUESTION=0x16;                 //回复问题
10    public static final int NO_DATA=-1024;                        //数据不存在
11    public static final String ERROR="error";                     //访问出错
12    public static final String EXCEPTION="exception";             //访问抛出异常
13  }
```

Global 类中定义一些全局变量，注意在 URL 常量中填入自己主机的 IP 地址，代码如程序清单 7-3 所示。

程序清单 7-3：Code0701\app\src\main\java\cn\jxufe\iet\code0701\util\Global.java

```
1   public class Global {                               //定义一些全局变量
2     public static int sourceAreaId=2;                 //生源地 Id,默认为江西
3     public static int targetAreaId=2;                 //意向省份 Id
4     public static int schoolAreaId=2;
5     public static int batchId=1;                      //批次 Id
6     public static int categoryId=1;                   //科类 Id
7     public static String DB_PATH;                     //数据库的路径
8     public static String DB_NAME="bkt.db";            //数据库的名称
9     public static String[] batches = new String[] { "不限", "一本", "二本", "三本",
10  "专科", "提前" };                                    //保存所有的批次,数组下标与其对应的 Id 一致
11    public static String[] categories = new String[] { "不限", "文史", "理工", "综合类",
12  "艺术类", "体育" };                                  //保存所有的类别,数组下标与其对应的 Id 一致
13    public static String URL="http://主机 IP 地址:8080/bktService/";
                                                        //本地服务器地址
14    public static DBUtil dbUtil;
15    public static SQLiteDatabase db;
16    public static SharedPreferences initPreferences;  //本地保存的信息
17  }
```

Util 工具类包含检查是否联网方法 isNetworkConnected 和获取手机当时时间方法 getDate，具体代码如程序清单 7-4 所示。

程序清单 7-4：Code0701\app\src\main\java\cn\jxufe\iet\code0701\util\Util.java

```
1   public class Util {
2       public static boolean isNetworkConnected(Context context) {
                                                        //判断网络是否连接
3           if (context != null) {
4               ConnectivityManager mConnectivityManager = (ConnectivityManager) context
5                       .getSystemService(Context.CONNECTIVITY_SERVICE);
                                                        //获取网络连接管理器
6               NetworkInfo mNetworkInfo;
7               mNetworkInfo = mConnectivityManager
8                       .getActiveNetworkInfo();
9               if (mNetworkInfo != null) {
10                  return mNetworkInfo.isAvailable();
11              }
12              Toast.makeText(context, "网络不可用,请检查你的网络状态!",
```

```
                 Toast.LENGTH_SHORT)
13                       .show();
14               Intent intent = null;
15               if (android.os.Build.VERSION.SDK_INT > 10) {//判断手机系统的版
                                                            //本,即API 大于10
16                   //就是3.0或以上版本
17                   intent = new Intent(
18                       android.provider.Settings.ACTION_WIRELESS_SETTINGS);
19               } else {
20                   intent = new Intent();
21                   ComponentName component = new ComponentName(
22  "com.android.settings",
23  "com.android.settings.WirelessSettings");
24                   intent.setComponent(component);
25                   intent.setAction("android.intent.action.VIEW");
26               }
27               context.startActivity(intent);
28           }
29           return false;
30       }
31       public static String getDate() {
32
33           Date date = new Date(System.currentTimeMillis());
34           SimpleDateFormat sdf = new SimpleDateFormat("yyyy-MM-dd HH:mm");
35           return sdf.format(date);
36       }
37  }
```

7.3 "省控线查询"模块与服务器端交互的实现

7.3.1 "省控线查询"模块与服务器端交互流程

在第4章中,省控线列表数据是从GetService类中获取的,临时代替从服务器端获取数据。本节主要实现监听省控线查询条件筛选栏获取查询参数,开启线程并调用AccessToServer类请求服务器端返回省控线列表JSON数据,在onCreate方法中解析JSON数据,实现从服务器端获取的数据以列表的形式显示。具体实现流程参见图7-12。

7.3.2 获取服务器端数据

监听省控线查询条件筛选栏获取查询参数在第4章中已经实现,本节中不用修改。之前省控线列表数据是从GetService类中获取的,现在需要在Controlline Fragment类中对这些获取临时数据的代码和列表显示代码进行注释,如图7-13所示。

在ControllineFragment类中定义Handler和Gson常量,代码如程序清单7-5所示。

第 7 章 App 客户端与服务器交互设计

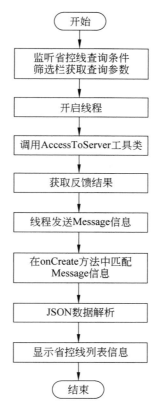

图 7-12 "省控线查询"模块实现的具体流程

```
/*显示省控线列标普*/
/*
GetService getService = new GetService();
controlLineList = getService.setControlLineList();//为controlLineList集合添加集合内容

ControlLineAdapter controlLineAdapter = new ControlLineAdapter
        (getActivity(), controlLineList);
controlLineListView.setAdapter(controlLineAdapter);*/
```

图 7-13 注释从 GetService 中获取数据的代码

程序清单 7-5：Code0701\app\src\main\java\cn\jxufe\iet\code0701\fragment\ControllineFragment.java

```
1    private Handler mHandler;                          //消息处理器
2    public Gson gson = new Gson();
3    private boolean isFinishedInit = false;            //是否完成初始化
```

定义 getControlLine 方法，在 getControlLine 中开启线程，调用 AccessToServer 方法获取服务器端反馈数据并发送消息标记，代码如程序清单 7-6 所示。

程序清单 7-6：Code0701\app\src\main\java\cn\jxufe\iet\ code0701\fragment\ControllineFragment.java

```java
1   public void getControlLine() {                              //查询省控线
2       DBUtil dbUtil=new DBUtil(getActivity());
3       final int area = dbUtil.getAreaId(getActivity(),
4               provinces.get(Global.sourceAreaId));            //获取省份编号
5       if (Util.isNetworkConnected(getActivity())) {           //判断网络是否有用
6           new Thread() {
7               public void run() {
8                   AccessToServer accessToServer = new AccessToServer();
9                   String result = accessToServer.doGet(Global.URL
10                          + "ControlLineServlet", new String[] { "year",
11  "batchId", "categoryId", "areaId" }, new String[] {
12                          years[yearId], batchId + "", categoryId + "",
13                          area + "" });
14                  System.out.println("result="+result);
15                  Message message = Message.obtain();
16                  message.what = Constants.GET_CONTROL_LINE;  //省控线结果消息标记
17                  message.obj = result;                       //省控线查询结果
18                  mHandler.sendMessage(message);              //发送消息
19              }
20          }.start();
21      }
22  }
```

7.3.3 显示省控线数据列表

getControlLine 方法编写完后，需要在其他方法中调用才能发挥它的作用。调用 getControlLine 方法有两个地方：一个是在 onCreateView 方法内调用；另一个位置是在 MyItemSelectedListener 事件监听类中，结合 isFinishedInit 判断是否初始化，然后调用 getControlLine 方法，如图 7-14 和图 7-15 所示。

```
MyItemSelectedListener itemSelectedListener = new MyItemSelectedListener();
yearSpinner.setOnItemSelectedListener(itemSelectedListener);
provinceSpinner.setOnItemSelectedListener(itemSelectedListener);
categorySpinner.setOnItemSelectedListener(itemSelectedListener);
batchSpinner.setOnItemSelectedListener(itemSelectedListener);

provinceSpinner.setSelection(sourceAreaId);
categorySpinner.setSelection(categoryId);
batchSpinner.setSelection(batchId);
getControlLine();
```

图 7-14 在 onCreateView 方法中调用 getControlLine 方法

从服务器端反馈 JSON 数据后，需要解析 JSON 数据，并以列表的形式显示。重写

```
            break;
        case R.id.batchSpinner:// 如果是选择批次列表
            batchId = position;
            break;
        default:
            break;
    }

    if (isFinishedInit) {
        getControlLine();
    }
}
```

图 7-15　在 MyItemSelectedListener 类中调用 getControlLine 方法

onCreate 方法，接收消息与省控线标识匹配成功后，解析 JSON 数据并以列表的形式显示，代码如程序清单 7-7 所示。

程序清单 7-7：Code0701\app\src\main\java\cn\jxufe\iet\code0701\fragment\ControllineFragment.java

```
1    @Override
2    public void onCreate(@Nullable Bundle savedInstanceState) {
3        super.onCreate(savedInstanceState);
4        mHandler = new Handler() {                      //处理请求结果
5
6            public void handleMessage(Message msg) {
7                String result = (String) msg.obj;
8                switch (msg.what) {
9                  case Constants.GET_CONTROL_LINE:
10                     if ("exception".equals(result) || "error".equals(result)
11                         || "".equals(result)) {
12                         Toast.makeText(getActivity(), "访问服务器异常!",
13                             Toast.LENGTH_SHORT);
14                         return;
15                     }
16                     controlLineList = gson.fromJson(result,
17                         new TypeToken<List<ControlLine>>() {
18                         }.getType());
19 if (controlLineList == null || controlLineList.size() == 0) {
20                     Toast.makeText(getActivity(), "数据维护中...请稍后再试!",
21                         Toast.LENGTH_LONG).show();
22                     } else {
23                         if (!isFinishedInit) {
24                             isFinishedInit = true;
25                         }                            //是否完成初始化操作
26                    ControlLineAdapter controlLineAdapter = new
27                        ControlLineAdapter(getActivity(), controlLineList);
```

```
28                    controlLineListView.setAdapter(controlLineAdapter);
29                }
30                break;
31            default:
32                break;
33        }
34    }
35  };
36 }
```

7.4 "历年录取线查询"模块与服务器端交互的实现

7.4.1 "历年录取线查询"模块与服务器端交互流程

"历年录取线查询"模块与服务器端交互的实现过程和"省控线查询"模块比较类似,主要是根据判断显示学校录取分数线和专业录取分数线。本节主要实现监听录取分数线查询条件筛选栏获取查询参数,根据判断开启学校录取分数线和专业录取分数线线程,调用AccessToServer类请求服务器端返回录取分数线列表JSON数据,在onCreate方法中解析JSON数据,实现从服务器端获取的数据以列表的形式显示。具体实现流程参见图7-16。

图7-16 "历年录取线查询"模块实现的具体流程

7.4.2 获取服务器端学校录取线和专业录取线

在 EnrollScoreFragment 类中定义 Handler 和 Gson 常量,代码如程序清单 7-8 所示。

程序清单 7-8：Code0702\app\src\main\java\cn\jxufe\iet\code0702\fragment\EnrollScoreFragment.java

```
1    private Handler mHandler;
2    private boolean isFinishedInit = false;       //是否完成初始化
3    private int schoolServiceId;                  //学校在服务器端数据库中的 Id
```

事件监听条件筛选栏发生一些改变,因为由原来本地临时数据转换成从服务器端获取数据,请求服务器端数据时需要传递被查询学校在服务器端的 Id 信息。因此在事件监听器类 MyItemSelectedListener 中作一些简单的调整,如图 7-17 所示。

```
private class MyItemSelectedListener implements AdapterView.OnItemSelectedListener {
    @Override
    public void onItemSelected(AdapterView<?> parent, View view,
                               int position, long id) {
        switch (parent.getId()) {
            case R.id.schoolAreaSpinner:// 如果选择学校所在地
                schoolAreaId = position;
                getSchool(provinces.get(schoolAreaId));// 当省份变化时,对应显示的学校都会随之变化
                break;
            case R.id.areaSpinner:// 如果是选择生源地
                sourceAreaId = position;
                break;
            case R.id.categorySpinner:// 如果是选择文理科
                categoryId = position+1;// 序号从0开始
                break;
            case R.id.batchSpinner:// 如果是选择批次
                batchId = position+1;
                break;
            case R.id.schoolSpinner:// 如果是选择学校
                schoolId = position;// 获取学校名称
                schoolServiceId = schoolList.get(position).getSchoolId();// 获取学校在服务器端数据库中的Id
                schoolName = schoolList.get(position).getSchoolName();// 获取学校名称
                break;
            case R.id.yearSpinner:
```

图 7-17 添加"获取学校在服务器端数据库中 id"代码

下面需要分别定义 getSchoolRecruits 和 getMajorRecruits 方法,分别启动线程调用 AccessToServer 方法获取服务器端反馈 JSON 数据并发送对应消息标记。

获取学校录取线的 getSchoolRecruits 方法的代码如程序清单 7-9 所示。

程序清单 7-9：Code0702\app\src\main\java\cn\jxufe\iet\code0702\fragment\EnrollScoreFragment.java

```
1    public void getSchoolRecruits() {                //获取学校录取线信息
2        DBUtil dbUtil=new DBUtil(getActivity());
3        final int area = dbUtil.getAreaId(getActivity(),
4                provinces.get(sourceAreaId));
5        new Thread() {
6            public void run() {
```

```
7            AccessToServer accessToServer = new AccessToServer();
8            String result = accessToServer.doGet(Global.URL
9                    + "SchoolRecruitsServlet", new String[]
   { "schoolId","categoryId", "batchId", "year",
10  "sourceAreaId" },
11                  new String[] { schoolServiceId + "", categoryId + "",
12                      batchId + "", years[yearId], area + "" });
13           Message message = Message.obtain();
14           message.obj = result;
15           message.what = Constants.GET_SCHOOL_ENROLL_SCORE;
16           mHandler.sendMessage(message);
17       }
18   }.start();
19 }
```

获取专业录取线的 getMajorRecruits 方法的代码如程序清单 7-10 所示。

程序清单 7-10：Code0702\app\src\main\java\cn\jxufe\iet\code0702\fragment\EnrollScoreFragment.java

```
1  public void getMajorRecruits() {                     //获取专业录取线
2      DBUtil dbUtil=new DBUtil(getActivity());
3      final int area = dbUtil.getAreaId(getActivity(),
4              provinces.get(sourceAreaId));
5      new Thread() {                                   //启动线程
6          public void run() {
7              AccessToServer accessToServer = new AccessToServer();
8              String result = accessToServer.doGet(Global.URL
9                      + "MajorRecruitServlet", new String[] { "schoolId",
10  "year", "areaId" }, new String[] { schoolServiceId + "", years[yearId],
11                      area + "" });             //发送请求并获取结果
12             Message message = Message.obtain();
13             message.obj = result;
14             message.what = Constants.GET_MAJOR_ENROLL_SCORE;
15             mHandler.sendMessage(message);
16         }
17     }.start();
18 }
```

7.4.3 显示学校录取线和专业录取线列表

数据获取后需要进行显示。在原来的项目中，是经过点击"查询"按钮后判断是否选择专业，直接调用数据显示方法显示对应数据。现在是先判断再从服务器端获取数据，获取服务器端数据后再显示数据。因此，需要对"查询"按钮的判断方法进行调整，由原来直接显示数据的方法转变成调用获取数据的方法，如图 7-18 所示。

从服务器端分别反馈学校录取线和专业录取线 JSON 数据后，需要解析 JSON 数据，

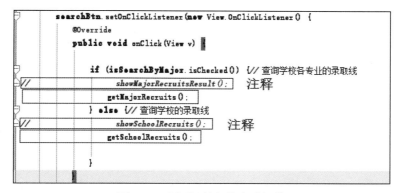

图 7-18 "查询"按钮的方法调整

并以列表的形式显示。重写 onCreate 方法,匹配是学校录取线线程发送信息或是专业录取线线程发送消息,并对 JSON 数据做相应解析,调用数据显示方法 showSchoolRecruits 和 showMajorRecruitsResult,以列表的形式显示,代码如程序清单 7-11 所示。

程序清单 7-11:Code0702\app\src\main\java\cn\jxufe\iet\code0702\fragment\EnrollScoreFragment.java

```
1   public void onCreate(@Nullable Bundle savedInstanceState) {
2       super.onCreate(savedInstanceState);
3       DBUtil dbUtil=new DBUtil(getActivity());
4       provinces = dbUtil.getField(getActivity(),
5   "select areaName  from area");                    //获取所有的省份
6       mHandler = new Handler() {
7           @Override
8           public void handleMessage(Message msg) {
9               String result = (String) msg.obj;
10              if ("error".equals(result) || "exception".equals(result)) {
11                  Toast.makeText(getActivity(), "访问服务器异常!",
12                          Toast.LENGTH_SHORT);
13                  return;
14              }
15              enrollResultView.setVisibility(View.VISIBLE);
16              switch (msg.what) {
17                  case Constants.GET_SCHOOL_ENROLL_SCORE:   //获取学校录取线
18                      schoolRecruits = gson.fromJson(result,
19                              new TypeToken<List<SchoolRecruit>>() {
20                              }.getType());
21                      if (schoolRecruits == null || schoolRecruits.size() == 0){
22                          Toast.makeText(getActivity(), "数据维护中...请稍后再试!",
23                                  Toast.LENGTH_LONG).show();
24                          enrollResultView.setVisibility(View.INVISIBLE);
25                      } else {                               //如果有数据,则显示查询结果
```

```
26                        showSchoolRecruits();        //显示学校录取数据
27                    }
28                    break;
29                case Constants.GET_MAJOR_ENROLL_SCORE:    //获取专业录取线
30                    majorRecruits = gson.fromJson(result,
31                        new TypeToken<List<MajorRecruit>>() {
32                        }.getType());
33                    if (majorRecruits == null || majorRecruits.size() == 0) {
34                      Toast.makeText(getActivity(), "数据维护中…请稍后再
35                                试!", Toast.LENGTH_LONG).show();
36                        enrollResultView.setVisibility(View.INVISIBLE);
37                    } else {                        //如果有数据,则显示查询结果
38                        showMajorRecruitsResult(); //显示专业录取数据
39                    }
40                    break;
41
42                default:
43                    break;
44            }
45        }
46    };
47 }
```

7.5 "报考咨询"模块与服务器端交互的实现

7.5.1 "报考咨询"模块与服务器端交互流程

在第4章中,"报考咨询"列表数据是从 GetService 类中获取的,临时代替从服务器端获取数据。本节主要实现打开本模块能够加载历史问题列表;在咨询问题编辑框中输入咨询问题后,点击"提交"按钮能够把咨询问题提交到服务器,并更新列表;在搜索文字框中输入查询内容后点击"查询"按钮,从服务器端获取查询数据;长按问题列表能够弹出对话框,输入回复内容后能够把内容更新到服务器端。具体的"报考咨询"模块与服务器端交互流程如图7-19所示。

7.5.2 获取历史问题列表

之前咨询问题列表数据是从 GetService 类中获取的,现在替换成从服务器端获取。由于代码量不是很大,因此现在需要把从 GetService 类中获取数据的代码、显示问题列表代码和列表长按事件监听代码注释掉,如图7-20所示。

从 GetService 类中获取临时数据的代码已经被注释,现在从服务器端获取历史数据。在 ConsultFragment 类中新建 getHistoryQuestion 方法,启动线程,调用 AccessToServer 方法获取服务器端反馈数据并发送消息标记,代码如程序清单7-12所示。

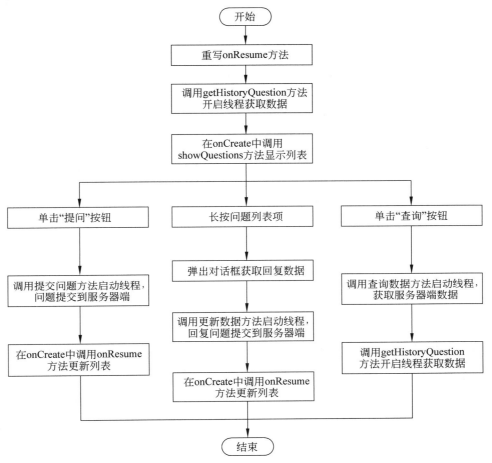

图 7-19 "报考咨询"模块实现的具体流程

```
//      GetService getService=new GetService();
//      questionList=getService.setQuestionList();
//
//      ConsultAdapter consultAdapter=new ConsultAdapter(getActivity(),questionList);
//      questionListView.setAdapter(consultAdapter);
//      questionListView.setOnItemLongClickListener(new AdapterView.OnItemLongClickListener() {
// 长按某一项进行回复
//          @Override
//          public boolean onItemLongClick(AdapterView<?> parent,
//                                  View view, int position, long id) {
//              updateQuestion();
//              return false;
//          }
//      });
```

图 7-20 将之前代码注释掉

程序清单 7-12：Code0703\app\src\main\java\cn\jxufe\iet\code0703\fragment\ConsultFragment.java

```
1    public void getHistoryQuestion(final String keyword) {//获取以往的问题
2        if (Util.isNetworkConnected(getActivity())) {
3            new Thread() {
4                public void run() {                              //线程执行体
5                    AccessToServer accessToServer = new AccessToServer();
6                    String result = accessToServer.doPost(Global.URL
7                            + "QuestionServlet", new String[] { "flag",
8  "keyword" },
9                            new String[] { Question.QUERY, keyword });
10                   System.out.println("result="+result);
11                   Message message = Message.obtain();
12                   message.obj = result;
13                   message.what = Constants.GET_QUESTIONS;
14                   mHandler.sendMessage(message);
15               }
16           }.start();
17       }
18   }
```

为达到打开本模块就能显示历史问题列表的目的，重写 onResume 方法，在 onResume 方法中调用 getHistoryQuestion 方法，具体请参见程序清单 7-13 所示的代码。

程序清单 7-13：Code0703\app\src\main\java\cn\jxufe\iet\code0703\fragment\ConsultFragment.java

```
1    @Override
2    public void onResume() {
3        super.onResume();
4        getHistoryQuestion(keywordView.getText().toString().trim());
5    }
```

从服务器端获取反馈的历史问题 JSON 数据后，需要解析 JSON 数据，并以列表的形式显示。重写 onCreate 方法，接收消息与获取问题标识匹配成功后，解析 JSON 数据，并调用 showQuestions 方法显示列表数据，代码如程序清单 7-14 所示。

程序清单 7-14：Code0703\app\src\main\java\cn\jxufe\iet\code0703\fragment\ConsultFragment.java

```
1    public void onCreate(Bundle savedInstanceState) {
2        super.onCreate(savedInstanceState);
3        mHandler = new Handler() {
4            @Override
5            public void handleMessage(Message msg) {
6                String result = (String) msg.obj;
7                result = (String) msg.obj;
```

```
8            switch (msg.what) {
9                case Constants.GET_QUESTIONS:
10                   if ("exception".equals(result) || "error".equals(result)
11                           || "".equals(result)) {
12                       Toast.makeText(getActivity(), "访问服务器异常!",
13                               Toast.LENGTH_SHORT);
14                       return;
15                   }
16                   questions = gson.fromJson(result,
17                           new TypeToken<List<Question>>() {
18                           }.getType());
19                   if (questions != null) {
20                       showQuestions();
21                   }
22                   break;
23               default:
24                   break;
25           }
26       }
27   };                                              //初始化 Handler 对象
28 }
```

在上面的代码中,我们发现 showQuestions 方法还没有实现。showQuestions 方法是显示问题列表,具体代码如程序清单 7-15 所示。

程序清单 7-15:Code0703\app\src\main\java\cn\jxufe\iet\code0703\fragment\ConsultFragment.java

```
1   public void showQuestions() {                    //显示问题列表
2       ConsultAdapter consultAdapter=new ConsultAdapter(getActivity(), questions);
3       questionListView.setAdapter(consultAdapter);
4   }
```

7.5.3 实现"提问"模块

点击"提问"按钮,能够把咨询编辑框内输入的内容提交到服务器端,并且能更新问题列表数据,如图 7-21 所示。首先需要监听"提问"按钮的匿名内部类事件,并把输入的问题封装成 JSON 数据格式,以便于后面提交到服务器端,具体代码如程序清单 7-16 所示。

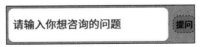

图 7-21 "提问"界面

程序清单 7-16：Code0703\app\src\main\java\cn\jxufe\iet\code0703\fragment\ConsultFragment.java

```java
1    submit.setOnClickListener(new View.OnClickListener() {//提交按钮的事件处理
2        public void onClick(View v) {
3            String questionContent = contentView.getText().toString()
4                    .trim();                                    //获取输入的问题
5            if ("".equals(questionContent)) {
6                AlertDialog.Builder builder = new AlertDialog.Builder(
7                        getActivity());
8                builder.setMessage("请输入咨询的问题,内容不能为空!");
                                                                 //对话框显示的内容
9                builder.create().show();                        //创建并显示对话框
10           } else {
11               Question question = new Question();
12               question.setQuestionContent(contentView.getText()
13                       .toString());
14               question.setQuestionTime(Util.getDate());
15               questionString = gson.toJson(question);
16               submitQuestion();
17           }
18       }
19   });                                                        //"提问"按钮的单击事件处理
```

获取提交问题并封装成 JSON 数据后，需要把封装数据提交到服务器端。在上面的代码中，我们发现 submitQuestion 方法还未实现，因此需要实现该方法，并通过调用 AccessToServer 工具类把数据提交到服务器端，具体代码如程序清单 7-17 所示。

程序清单 7-17：Code0703\app\src\main\ java\cn\jxufe\iet\code0703\fragment\ConsultFragment.java

```java
1    public void submitQuestion() {
2        if (Util.isNetworkConnected(getActivity())) {
3            new Thread() {
4                public void run() {                             //线程执行体
5                    AccessToServer accessToServer = new AccessToServer();
6                    String result = accessToServer.doPost(Global.URL
7                            + "QuestionServlet", new String[] { "question",
8    "flag" }, new String[] { questionString,
9                            Question.INSERT });
10                   Message message = Message.obtain();
11                   message.obj = result;
12                   message.what = Constants.SUBMIT_QUESTION;
13                   mHandler.sendMessage(message);
14               }
15           }.start();
```

```
16      }
17  }
```

问题提交到服务器端后,需要更新问题列表数据。在 onCreate 方法的 switch 判断中添加"case Constants.SUBMIT_QUESTION:",对 result 信息进行匹配,再做进一步处理,具体代码如程序清单 7-18 所示。

程序清单 7-18:Code0703\app\src\main\java\cn\jxufe\iet\code0703\fragment\ConsultFragment.java

```
1   case Constants.SUBMIT_QUESTION:
2       if ("exception".equals(result) || "error".equals(result)
3               || "".equals(result)) {
4           Toast.makeText(getActivity(), "访问服务器异常!",
5                   Toast.LENGTH_SHORT);
6           return;
7       }
8       if ("true".equals(result)) {                //提交成功
9           Toast.makeText(getActivity(), "问题提交成功,请等待回复!",
10                  Toast.LENGTH_LONG).show();
11          contentView.setText("");
12          onResume();
13      } else {                                    //提交失败
14          Toast.makeText(getActivity(), "问题提交失败,请重新提交!",
15                  Toast.LENGTH_LONG).show();
16      }
17      break;
```

7.5.4 实现"问题回复"模块

长按问题列表后会弹出"我来回答"对话框,获取回复信息,如图 7-22 所示。点击"确定"按钮后,问题将会提交到服务器端,然后更新问题列表。首先需要对列表进行长按匿名内部类事件监听,调用更新问题对话框的 updateQuestion 方法,具体代码如程序清单 7-19 所示。

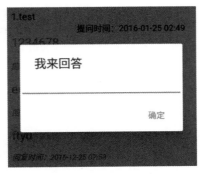

图 7-22 "问题回复"界面

程序清单 7-19：Code0703\app\src\main\java\cn\jxufe\iet\code0703\fragment\ConsultFragment.java

```
1   questionListView.setOnItemLongClickListener(new AdapterView
    .OnItemLongClickListener() {
2       //长按某一项进行回复
3       @Override
4       public boolean onItemLongClick(AdapterView<?> parent,
5                                     View view, int position, long id) {
6           updateQuestion(questions.get(position));
7           return false;
8       }
9   });
```

在 updateQuestion 方法中实现弹出对话框，获取输入内容，并调用 Util 类中的 getDate 方法，获取本地时间。然后把数据封装到 question 实体类中，具体代码如程序清单 7-20 所示。

程序清单 7-20：Code0703\app\src\main\java\cn\jxufe\iet\code0703\fragment\ConsultFragment.java

```
1   public void updateQuestion(final Question question) {
                                                            //更新问题状态
2       AlertDialog.Builder builder = new AlertDialog.Builder(getActivity());
                                                            //实例化 Builder 对象
3       builder.setTitle("我来回答");
4       final EditText answerText = new EditText(getActivity());
                                                            //实例化 EditText 对象
5       builder.setView(answerText);        //把编辑框以 View 的形式加入对话框
6       builder.setPositiveButton("确定", new DialogInterface.OnClickListener() {
                                                            //设置"确定"按钮
7           @Override
8           public void onClick(DialogInterface dialog, int which) {
9               String answer = answerText.getText().toString();
10              if (!"".equals(answer)) {    //如果回答不为空,则更新数据库
11                  String oldAnswer = question.getAnswerContent();
12                  if (oldAnswer != null && !"null".equals(oldAnswer)) {
13                      oldAnswer += "#" + answer;
14                  } else {
15                      oldAnswer = answer;
16                  }
17                  question.setAnswerContent(oldAnswer);
18                  String oldAnswerTime = question.getAnswerTime();
19                  if (oldAnswerTime != null && !"null".equals(oldAnswerTime)) {
20                      oldAnswerTime += "#" + Util.getDate();
21                  } else {
```

```
22                    oldAnswerTime = Util.getDate();
23                }
24                question.setAnswerTime(oldAnswerTime);
25                update(question);
26            }
27        }
28    });
29    builder.create().show();                        //创建对话框
30 }
```

上述代码中的 update 方法还未实现,而回复内容和回复时间添加到 question 实体对象后,需要通过 update 方法调用 AccessToServer 工具类,把数据提交到服务器端,具体代码如程序清单 7-21 所示。

程序清单 7-21：Code0703\app\src\main\java\cn\jxufe\iet\code0703\fragment\ConsultFragment.java

```
1  public void update(Question question) {
2      Gson gson = new Gson();
3      questionString = gson.toJson(question);
4      if (Util.isNetworkConnected(getActivity())) {
5          new Thread() {
6              public void run() {                    //线程执行体
7                  AccessToServer accessToServer = new AccessToServer();
8                  String result = accessToServer.doPost(Global.URL
9                          + "QuestionServlet", new String[] { "question",
10  "flag" }, new String[] { questionString,
11                          Question.UPDATE });
12                 Message message = Message.obtain();
13                 message.obj = result;
14                 message.what = Constants.UPDATE_QUESTION;
15                 mHandler.sendMessage(message);
16             }
17         }.start();
18     }
19 }
```

问题提交到服务器端后,需要更新问题列表数据。在 onCreate 的 switch 判断方法中添加"case Constants.UPDATE_QUESTION：",对 result 信息进行匹配后再做进一步处理,具体代码如程序清单 7-22 所示。

程序清单 7-22：Code0703\app\src\main\java\cn\jxufe\iet\code0703\fragment\ConsultFragment.java

```
1  case Constants.UPDATE_QUESTION:
2      if ("exception".equals(result) || "error".equals(result)
3              || "".equals(result)) {
```

```
 4              Toast.makeText(getActivity(), "访问服务器异常!",
 5                    Toast.LENGTH_SHORT);
 6              return;
 7          }
 8          if ("true".equals(result)) {                    //提交成功
 9              Toast.makeText(getActivity(), "回复成功,请等待审核",
10                    Toast.LENGTH_LONG).show();
11              onResume();
12          } else {                                        //提交失败
13              Toast.makeText(getActivity(), "回复失败,请重试!",
14                    Toast.LENGTH_LONG).show();
15          }
16          break;
```

7.5.5 实现"查询问题"模块

点击"查询"按钮能够把查询编辑框内输入的内容提交到服务器端,并且能更新问题列表数据,如图 7-23 所示。首先需要监听"查询"按钮的匿名内部类事件,获取搜索内容,将搜索内容作为参数调用 getHistoryQuestion 方法,具体代码如程序清单 7-23 所示。

图 7-23 "查询问题"界面

程序清单 7-23：Code0703\app\src\main\java\cn\jxufe\iet\code0703\fragment\ConsultFragment.java

```
1   search.setOnClickListener(new View.OnClickListener() {
                                                    //"查询"按钮的单击事件处理
2       @Override
3       public void onClick(View v) {
4           getHistoryQuestion(keywordView.getText().toString().trim());
5       }
6   });
```

getHistoryQuestion 方法前面已经实现,这里直接调用即可。在 getHistoryQuestion 方法中也会启动线程,通过 AccessToServer 工具类根据搜索条件筛选服务器端问题列表,以 JSON 封装形式反馈数据。在线程中发送消息后,在 onCreate 方法中解析 JSON 数据,并以列表显示,相关代码前面已经提供,在此不再提供。

7.6 本章小结

本章主要讲解了 Android 客户端与服务器端之间进行数据的传递和交互,Android 提供的多种访问网络的形式和相关的 API;介绍了如何通过 Apache HttpClient 发送请求和获取服务器端的处理结果，涉及 HttpClient、HttpGet、HttpPost、HttpEntity、HttpResponse 等类的知识。在客户端发送请求时，需要注意在 Android 2.3 以后不允许直接访问网络，建议采用多线程访问网络。

本章使用 HttpClient、JavaBean、多线程、Handler、JDBC、MySQL 等相关知识，实现了"豹考通"服务器的"省控线查询"模块"历年录取线查询"模块"报考咨询"模块客户端与服务器端的交互，在客户端使用列表展示结果。

7.7 课后练习

1. 使用 WebView 控件显示百度首页。
2. 使用 HTTP 发送请求有哪几种方式？有何特点？
3. 简要描述 HTTP 中 Get 和 Post 的区别。
4. 简要描述 HttpClient 发送请求获取结果的执行过程。
5. 简要描述 JSON 封装和解析的过程。
6. 制作一个程序，程序的主页面上有一个 Button 和一个 TextView。点击 Button 后会向指定地址请求数据，将这些数据以字符串的形式显示在 TextView 上（地址为 http://www.drhelp.cn/bkt/ControlLineServlet?areaId=13&year=2011）。
7. 在第 6 题的基础上，将 TextView 改成 ListView 并解析请求到的 JSON 数据，将数据按照规则显示在 ListView 上（要求每个省控线实体逐行显示）。

第 8 章

Spring Boot 服务器端设计

8.1 本章简介

对于 App 而言,服务器可以简单理解为 App 的后台,而作为 App 的后台,则需要与数据库进行交互,并且能够满足用户对于数据库的增、删、改、查等一系列操作要求。

本章将以当下较为流行的 Spring Boot+Mybatis 框架为基础,讲述通过此框架进行"豹考通"App 后台服务器的搭建与对"豹考通"教学案例数据库的业务逻辑操作,并且结合"豹考通"教学案例,让大家深刻理解 Spring Boot+Mybatis 框架的优势,同时熟练掌握该框架的基本内容。

8.2 Spring Boot 开发基础

8.2.1 Spring Boot 技术简介

多年以来,Spring 平台饱受非议的一点就是大量的 XML 配置以及复杂的依赖管理。

随着使用 Spring 进行开发的个人和企业越来越多,Spring 也慢慢从一个单一简洁的小框架变成一个大而全的开源软件。Spring 的边界不断扩充,到了后来它几乎可以做任何事情,市面上主流的开源软件和中间件都有 Spring 对应组件支持,人们在享用 Spring 的便利之后也遇到了一些问题。

Spring 每集成一个开源软件,就需要增加一些基础配置,随着开发项目的逐渐庞大,往往需要集成很多开源软件。后期使用 Spring 开发大型项目需要引入很多配置文件,导致配置工作难以理解且出错率高,以至于人们称 Spring 为配置地狱。

在 2013 年的 SpringOne2GX 会议上,Pivotal 的首席技术官 Adrian Colyer 回应了这些批评,并且特别提到该平台将来的目标之一就是实现免 XML 配置的开发体验。Spring Boot 所实现的功能超出了这个任务的描述,开发人员不仅不再需要编写 XML,而且在一些场景中甚至不需要编写烦琐的 import 语句。

2013 年,微服务的概念也慢慢兴起,快速开发微小独立的应用变得更为急迫。Spring 刚好处在这样一个交叉点上,Pivotal 团队于 2013 年初启动了 Spring Boot 项目的研发。2014 年,Spring Boot 伴随着 Spring 4.0 诞生发布了第一个正式版本。

Spring Boot 并不是要成为 Spring 平台中众多 Foundation 层项目的替代者。它的目标不在于为已解决的问题域提供新的解决方案,而是为平台带来另一种开发体验,从而简化对这些已有技术的使用。对于已经熟悉 Spring 生态系统的开发人员来说,Spring Boot

是一个很理想的选择；对于采用 Spring 技术的新人来说，Spring Boot 提供一种更简洁的方式来使用这些技术。

Spring Boot 是由 Pivotal 团队提供的全新框架，其设计目的是简化 Spring 应用的初始搭建以及开发过程。该框架使用了特定的方式进行配置，从而使开发人员不再需要定义样板化的配置。Spring Boot 其实就是一个整合了很多可插拔组件（框架），内嵌了使用工具（例如 Tomcat、Jetty 等），且方便开发人员快速搭建和开发的一个框架。

8.2.2　Spring Boot 项目开发环境

1. Maven 项目管理工具的安装与配置

Maven 是基于项目对象模型（POM），利用一个中央信息片断管理一个项目的构建、报告和文档等。

（1）下载安装。从 Maven 官网（https://maven.apache.org/）下载 Apache Maven 3.6.3，其中 Windows 操作系统的用户选择 apache-maven-3.6.3-bin.zip；Linux 和 macOS 操作系统的用户选择 apache-maven-3.6.3-bin.tar.gz，具体如图 8-1 所示。

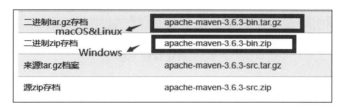

图 8-1　下载安装 Maven

（2）配置 Maven 环境变量。下载并安装成功之后，右击"计算机"图标，选择"属性"选项，之后单击"高级系统设置"→"环境变量"来设置环境变量。

如图 8-2 所示，新建系统变量 MAVEN_HOME，变量值为 Maven 的安装目录，此处为 D:\Program Files\Maven\apache-maven-3.6.3。编辑系统变量 Path，添加变量值%MAVEN_HOME%\bin，如图 8-3 所示。

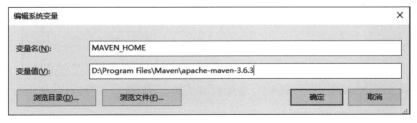

图 8-2　配置环境变量

注意，多个值之间需要有分号隔开，然后单击"确定"按钮。

2. Spring Boot 项目搭建

首先打开 IntelliJ IDEA 编译器，单击 Creat New Project 创建项目，如图 8-4 所示。选择 Spring Initializr 服务地址与相应的 JDK 版本后单击 Next 按钮，如图 8-5 所示。

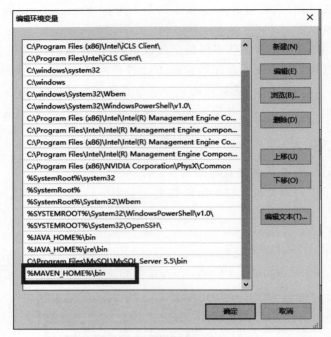

图 8-3　编辑环境变量

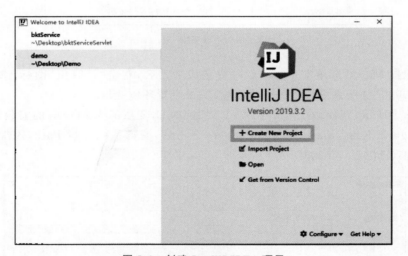

图 8-4　创建 IntelliJ IDEA 项目

如图 8-6 所示，输入 Project Metadata 的相关信息。需要注意的是，Artifact 中不能输入大写字母，可以选择 JAR 包或 WAR 包。由于 Spring Boot 已集成 Tomcat，因此可以直接用 java -jar name.jar 方式运行。

单击 Next 按钮后，选择需要添加的依赖，此处选择 Spring Web，单击 Next 按钮，如图 8-7 所示。

如图 8-8 所示，选择项目的存放目录，单击 Finish 按钮。

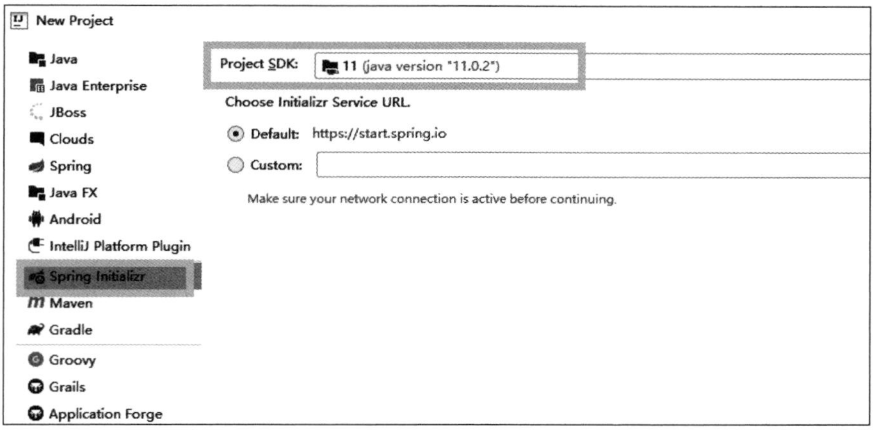

图 8-5　选择 Spring Initializr 服务地址

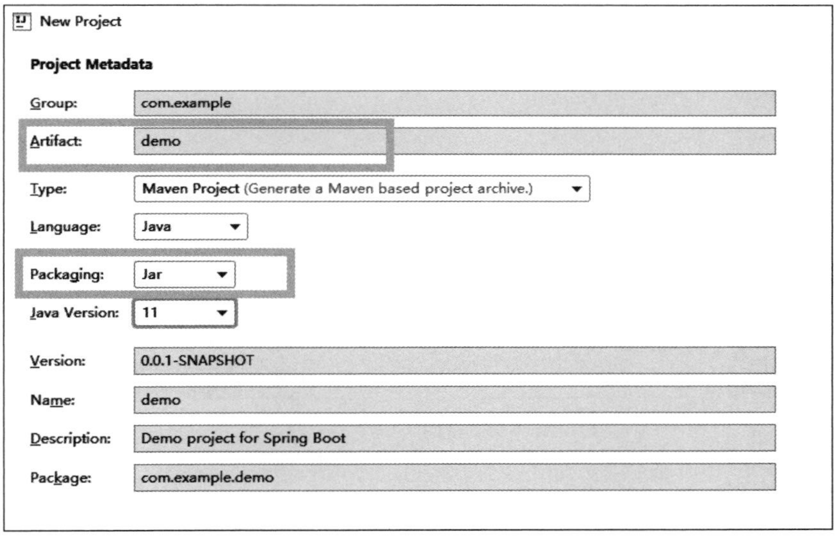

图 8-6　Project Metadata 相关配置

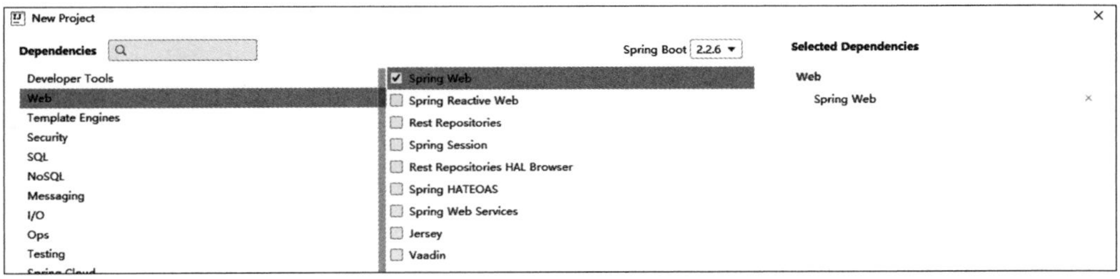

图 8-7　Spring Web 项目配置

等待依赖下载完成，Spring Boot 项目便创建完成。

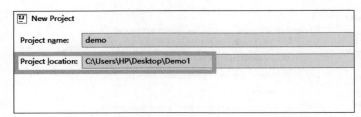

图 8-8　Spring Boot 项目路径

8.2.3　Spring Boot 项目开发基本过程

1. 通过注解处理网络请求

如图 8-9 所示，右击 com.example.demo 目录，选择 New 选项，然后单击 Package，创建 Controller 包。

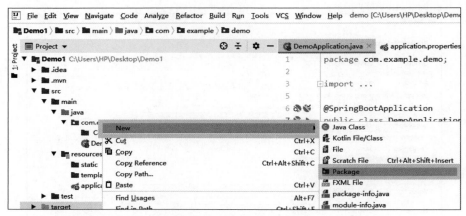

图 8-9　新建 Controller 包

创建完成后的结果如图 8-10 所示。

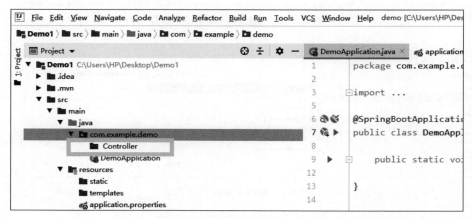

图 8-10　成功创建 Controller 包

然后，右击 Controller 包，选择 New 选项，单击 Java Class。将文件命名为 TestController，如图 8-11 所示。

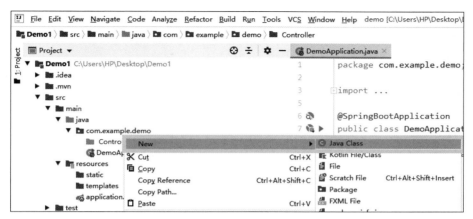

图 8-11　创建控制器

在 TestController 类中创建方法 Test()，返回字符串"Test Succeed!"作为测试成功的返回值，如图 8-12 所示。

```
public class TestController {
    public String Test(){
        return "Test Succeed!";
    }
}
```

图 8-12　Test 测试方法

之后，在 TestController 类中添加注解@RestController，告诉 Spring Boot 此类作为控制器；并且在方法 Test() 上方添加注解@RequestMapping，告诉 Servlet 我们的请求路径，其中路径名为@RequestMapping 注解括住的内容，此处我们以/hello 为路径名，如图 8-13 所示。

```
import org.springframework.web.bind.annotation.RequestMapping;
import org.springframework.web.bind.annotation.RestController;

@RestController
public class TestController {
    @RequestMapping("/hello")
    public String Test(){
        return "Test Succeed!";
    }
}
```

图 8-13　TestController 测试类

最后，如图 8-14 所示，找到 DemoApplication 文件，右击 Run DemoApplication。

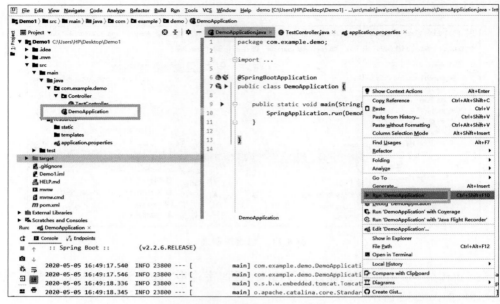

图 8-14 项目启动方式

控制栏出现图 8-15 所示页面，则表示连接成功。

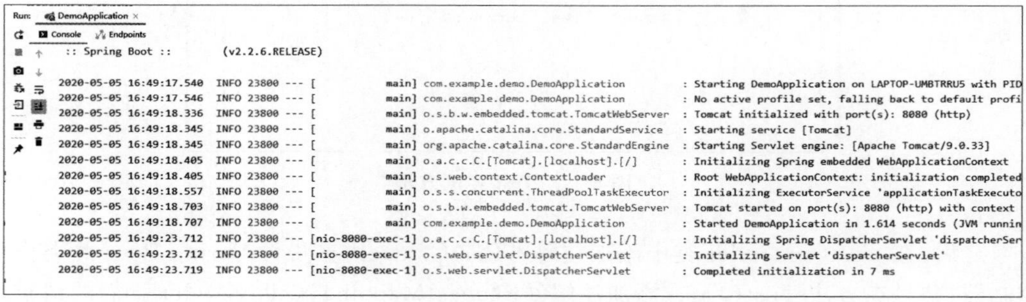

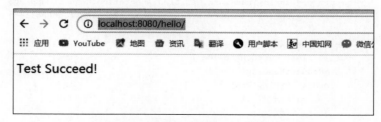

图 8-15 项目启动成功的控制台页面

此时，我们打开浏览器，输入 http://localhost:8080/hello/，出现如图 8-16 所示的页面，则表示网络请求处理成功。

图 8-16 项目成功启动的浏览器页面

2. 通过配置文件实现数据库的连接配置

如图 8-17 所示,打开项目目录,找到 application.properties 文件。此处为了方便后面的项目配置,我们右击该文件,选择 Refactor 选项,之后再选择 Rename 选项,将该文件重命名为 application.yml。

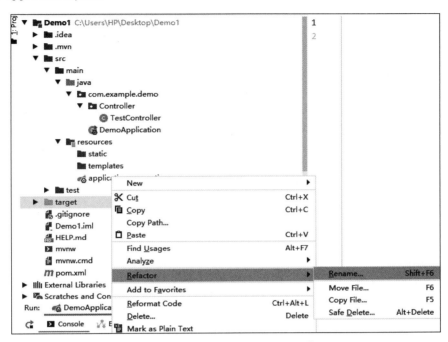

图 8-17 创建 application.yml 文件

然后打开文件,按照程序清单 8-1 所示输入代码。

程序清单 8-1:Code0801\Demo1\src\main\resources\application.yml

```
1    spring:
2      datasource:
3        url: jdbc:mysql://localhost:3306/bkt?useUnicode=true&characterEncoding=UTF-8&useSSL=false
4        username: root
5        password: 123456
6        driver-class-name: com.mysql.cj.jdbc.Driver
7    spring:
8      datasource:
```

其中,url 代表数据库的链接地址;username 为数据库用户名,password 为密码;driver-class-name 为驱动名称。

注意,此时驱动名称之所以会标红,是因为我们没有在 Maven 的 pom.xml 配置文件中添加 mysql 数据库依赖。我们需要打开项目目录,打开 pom.xml 配置文件,向其中添加如程序清单 8-2 所示的依赖。

程序清单 8-2：Code0801\Demo1\pom.xml

```
1    <dependency>
2        <groupId>mysql</groupId>
3        <artifactId>mysql-connector-java</artifactId>
4        <scope>runtime</scope>
5    </dependency>
```

3. 使用 Mybatis 对数据库进行管理

（1）导入依赖关系并配置 mybatis 文件。找到项目目录的 pom.xml 文件，单击进入，添加如程序清单 8-3 所示的依赖。

程序清单 8-3：Code0801\Demo1\pom.xml

```
1    <dependency>
2        <groupId>log4j</groupId>
3        <artifactId>log4j</artifactId>
4        <version>1.2.17</version>
5    </dependency>
6    <dependency>
7        <groupId>mysql</groupId>
8        <artifactId>mysql-connector-java</artifactId>
9        <version>5.1.46</version>
10   </dependency>
11   <dependency>
12       <groupId>org.mybatis.spring.boot</groupId>
13       <artifactId>mybatis-spring-boot-starter</artifactId>
14       <version>1.2.0</version>
15   </dependency>
```

之后，进入 application.yml 配置文件，添加 mybatis 的映射与实体路径的配置语句，如程序清单 8-4 所示。

程序清单 8-4：Code0801\Demo1\src\main\resources\application.yml

```
1    mybatis:
2      mapper-locations: classpath:mapper/*Mapper.xml
3      type-aliases-package: com.example.demo.bean
```

（2）配置 mybatis-generator.xml 文件。进入项目目录的 src/resources 文件夹，通过右击新建文件，命名为 generatorConfig.xml，输入如程序清单 8-5 所示的配置代码。

程序清单 8-5：Code0801\Demo1\src\main\resources\generatorConfig.xml

```
1    <?xml version="1.0" encoding="UTF-8"?>
```

```
2    <!DOCTYPE generatorConfiguration
3            PUBLIC "-//mybatis.org//DTD MyBatis Generator Configuration 1.0//EN"
4            "http://mybatis.org/dtd/mybatis-generator-config_1_0.dtd">
5    <generatorConfiguration>
6        <!--classPathEntry location="D:\zngkpt\m2\repository\mysql\mysql-connector-java\5.1.40\mysql-connector-java-5.1.40.jar" /-->
7        <classPathEntry
8        location="C:\Users\HP\Desktop\Demo1\mysql-connector-java-5.1.6-bin.jar"/>
9        <context id="context1" targetRuntime="MyBatis3">
10           <commentGenerator>
11           <!-- 去除自动生成的注释 -->
12           <property name="suppressAllComments" value="true" />
13           </commentGenerator>
14           <!-- 数据库连接配置 -->
15           <jdbcConnection driverClass="com.mysql.jdbc.Driver"
16           connectionURL="jdbc:mysql://localhost:3306/数据库名称"
17           userId="数据库用户名"
18           password="数据库连接密码" />
19           <!--非必需,类型处理器,在数据库类型和Java类型之间的转换控制-->
20           <javaTypeResolver>
21           <property name="forceBigDecimals" value="false"/>
22           </javaTypeResolver>
23           <!--配置生成的实体包
24           targetPackage:生成的实体包位置,默认存放在src目录下
25           targetProject:目标项目名-->
26           <javaModelGenerator targetPackage="com.example.demo.bean"
27           targetProject="src/main/java" />
28           <!-- 实体包对应映射文件位置及名称,默认存放在src目录下 -->
29           <sqlMapGenerator targetPackage="mapper"
30           targetProject="src/main/resources" />
31           <javaClientGenerator targetPackage="com.example.demo.dao"
32           targetProject="src/main/java" type="XMLMAPPER"/>
33            <!-- 配置表 schema:不用填写 tableName:表名 enableCountByExample、
     enableSelectByExampleenableDeleteByExampleenableUpdateByExample、
     selectByExampleQueryId:去除自动生成的例子-->
34           <table schema="" tableName="teacher" enableCountByExample="false"
     enableSelectByExample="false"
35              enableDeleteByExample = " false" enableUpdateByExample = " false"
     selectByExampleQueryId="false">
36           </table>
37       </context>
38   </generatorConfiguration>
```

（3）逆向建立项目的bean、dao、mapper文件。首先进入pom.xml文件,添加如程序

清单 8-6 所示的依赖。

程序清单 8-6：Code0801\Demo1\pom.xml

```
1   <dependency>
2       <groupId>org.mybatis.generator</groupId>
3       <artifactId>mybatis-generator-core</artifactId>
4       <version>1.3.6</version>
5   </dependency>
```

然后在＜build＞标签内添加如程序清单 8-7 所示的插件。

程序清单 8-7：Code0801\Demo1\pom.xml

```
1   <plugin>
2       <groupId>org.mybatis.generator</groupId>
3       <artifactId>mybatis-generator-maven-plugin</artifactId>
4       <version>1.3.6</version>
5       <!--在 pom.xml 中,在配置 MBG 插件时,可以通过 configuration
6       标签指定 MBG 的配置文件名、是否覆盖同名文件、是否将生成过程输出至控制台等-->
7       <configuration>
8           <!--mybatis-generator 的配置文件 generatorconfig.xml 的位置-->
9           <configurationFile>src/main/resources/generatorConfig.xml</configurationFile>
10          <!--是否将生成过程输出至控制台-->
11          <verbose>true</verbose>
12          <!--是否覆盖同名文件(只是针对 XML 文件,Java 文件生成类似 *.java.1、*.java.2 形式的文件)-->
13          <overwrite>true</overwrite>
14      </configuration>
15  </plugin>
16  <plugin>
```

最后，如图 8-18 和图 8-19 所示，单击右上角的 Edit Configuration 选项，进入选择界面。选择左上角的＋号按钮，选择添加 Maven 项目。

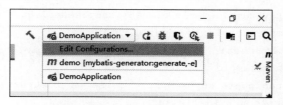

图 8-18　选择 Edit Configuration 选项

如图 8-20 所示，在命令行内输入 mybatis-generator:generate -e，运行该 Maven 项目。

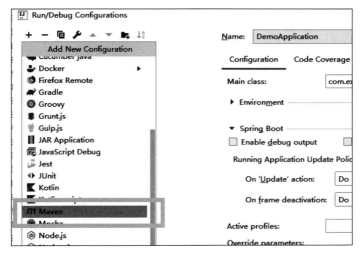

图 8-19　选择添加 Maven 项目

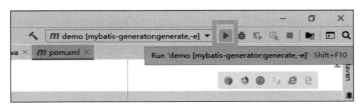

图 8-20　启动 Maven 项目

打开项目目录，如果出现如图 8-21 所示的包与类，则表示项目成功启动。

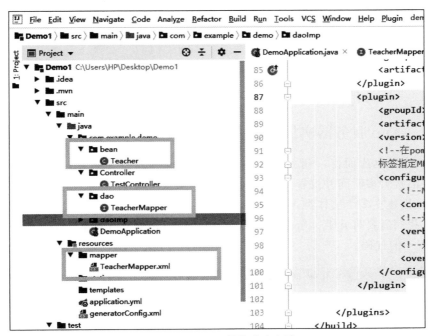

图 8-21　Maven 项目成功启动

8.3 App 服务器实体层设计与实现

8.3.1 App 服务器实体层设计

在本章采用的"豹考通"教学案例中,对后台数据库的操作与管理功能的实现主要通过 Bean(实体层,也称 POJO、Entity)、Dao(数据持久层,也称 Mapper)、Service(业务逻辑层)、Controller(控制层)这四层之间的相互协作来完成。

下面先介绍 Bean。该层主要是将数据库的表与 Java 代码中的类进行一一对应,也就是说,每一个 Bean 对应着数据库中的一张表,而每一个实体类中的成员变量对应着每一个表中的一个字段。以图 8-22 所示数据库中 area 这张表为例子详细展开。

图 8-22　area 数据表

如图 8-23 所示,对于这张表来说,存在两个字段,那么在其对应的 Bean 实体类中,也应该有以这两个字段命名的成员变量(areaId 和 areaName),其数据类型应与数据库表中的类型一致。

图 8-23　Area 类的成员属性

8.3.2 App 服务器实体层实现

对于一个数据库而言,一般情况下肯定存在多张表,如果对每张表分别手动建立一个实体类,会浪费大多时间并且容易出错。Mybatis 框架为我们提供了逆向工程工具,可以帮助我们轻松建立多个实体类,具体实现过程如下。

首先,我们需要打开上一节中创建的 generatorConfig.xml 配置文件,并找到如程序清单 8-8 所示的代码。

程序清单 8-8：Code0801\Demo1\src\main\resources\generatorConfig.xml

```
1    <table schema="" tableName="teacher" enableCountByExample="false"
2        enableSelectByExample="false" enableDeleteByExample
```

```
3        ="false" enableUpdateByExample="false"
4        selectByExampleQueryId="false">
5    </table>
```

将其中的 tableName="teacher" 替换为 tableName="表名"，例如 tableName="area"。注意，这里可以同时添加多个＜table＞标签，但每一个标签只能对应一张表，并且表名一定要与数据库的表名相对应，具体如图 8-24 所示。

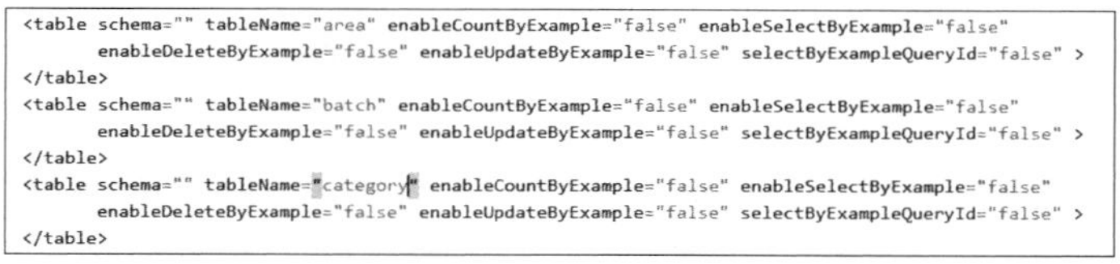

图 8-24　generatorConfig.xml 配置文件

最后，如图 8-25 所示，单击右上角的启动按钮，运行 Maven 项目。

图 8-25　启动配置文件

项目目录中出现图 8-26 所示的类与包，则表示成功构建实体层。

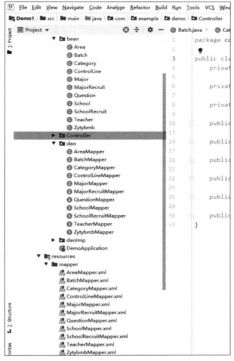

图 8-26　成功构建实体层

8.4　App 服务器数据持久层设计与实现

8.4.1　App 服务器数据持久层设计

数据持久层位于领域层和基础架构层之间。由于对象范例和关系范例这两大领域之间存在"阻抗不匹配",因此把数据持久层单独作为 J2EE 体系的一个层提出来,就能够在对象-关系数据库之间提供一个成功的企业级映射解决方案,尽最大可能弥补这两种范例之间的差异。

在 Mybatis 框架中,通过"Mapper 接口＋Mapper 映射"来实现 Dao。

根据上面的介绍,我们可以清晰地看到,每一张数据库的表都对应一个"Mapper 接口＋Mapper 映射"。因此,我们的逆向工程文件在创建 Bean 时,也自动创建每张表对应的"Mapper 接口＋Mapper 映射",如图 8-27 所示。

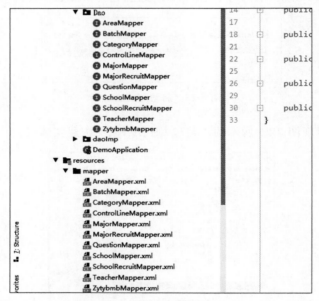

图 8-27　项目接口和映射目录

8.4.2　App 服务器数据持久层实现

1. "豹考通"数据库基本操作语句

对于数据库而言,存在的基本操作是对数据库单表的增、删、改、查。由于"豹考通"App 后台服务器涉及的表比较多,因此本节通过表 area 来进行案例讲解。

首先打开 AreaMapper.java 文件,可以看到 Mybatis 框架为我们创建的不是一个简单的类,而是一个接口。

通过 Java 的学习可以知道,接口类似于一种协议,例如两个开发者的开发时间完全不一致,那么需要两个人配合开发,由一个人将接口写好,定义好其中所有的变量命名规范和函数定义规范。具体实现类的开发人员只需要按照接口实现相应功能即可。因此，Mybatis 框架根据对数据库的基本操作需求（增删改查），为我们自动创建了接口中存在的基本操作方法，如表 8-1 所示。

表 8-1　接口方法与作用对应关系表

方　　法	作　　用
deleteByPrimaryKey(Short areaid)	根据主键 Id 删除
insert(Area record)	插入整条数据
insertSelective(Area record)	插入非空数据
selectByPrimaryKey(Short areaid)	通过主键查询信息
updateByPrimaryKeySelective(Area record)	对字段进行判断再更新（如果为 Null 就忽略更新）
updateByPrimaryKey(Area record)	对注入的字段全部更新

Mybatis 框架不仅为我们提供了这些基本的操作，还创建了这些基本操作对应的映射文件。此时需要打开 src/resource/mapper/AreaMapper.xml 文件，之后可以看见，Mybatis 已经为我们配置好映射文件。需要注意以下几个部分，首先是程序清单 8-9 和程序清单 8-10。

程序清单 8-9：Code0801\Demo1\src\resource\mapper\AreaMapper.xml

```
1    <mapper namespace="com.example.demo.dao.AreaMapper">
```

程序清单 8-10：Code0801\Demo1\src\resource\mapper\AreaMapper.xml

```
1    <resultMap id="BaseResultMap" type="com.example.demo.bean.Area">
2        <id column="areaId" jdbcType="SMALLINT" property="areaid" />
3        <result column="areaName" jdbcType="VARCHAR" property="areaname" />
4    </resultMap>
```

在 namespace＝"com.example.demo.dao.AreaMapper"这一句中，我们看到 com.example.demo.dao.AreaMapper 对应的就是 Dao 中的 AreaMapper 接口类，因此此句的作用便是将该映射与对应的类进行绑定。

＜resultMap＞标签表示将查询结果集中的列一一映射到 Bean 对象的各个属性。映射的查询结果集中的列标签可以根据需要灵活变化，并且在映射关系中，还可以通过 TypeHandler 设置实现查询结果值的类型转换，例如布尔型与 0/1 的类型转换。

＜id＞和＜result＞标签中的 column 表示数据库中表的字段（一定要与数据库中表的字段一一对应），property 表示其对应 Bean 中的成员变量的名称，一定要与 Bean 中成员变量名称一一对应。

其次，我们要看的是如程序清单 8-11 所示的查询语句。

程序清单 8-11：Code0801\Demo1\src\resource\mapper\AreaMapper.xml

```
1   <select
        id="selectByPrimaryKey"
        parameterType="java.lang.Short"
    resultMap="BaseResultMap">
2       select
3   <include refid="Base_Column_List" />
4       from area
5       where areaId = #{areaid,jdbcType=SMALLINT}
6   </select>
7   <delete id="deleteByPrimaryKey" parameterType="java.lang.Short">
8       delete from area
9       where areaId = #{areaid,jdbcType=SMALLINT}
10  </delete>
11  <insert id="insert" parameterType="com.example.demo.bean.Area">
12      insert into area (areaId, areaName)
13      Values(#{areaid,jdbcType=SMALLINT}, #{areaname,jdbcType=VARCHAR})
14  </insert>
```

根据上面的代码可以看出，所有的查询语句都是按照标签的形式书写。其中需要注意的是，id="selectByPrimaryKey"这一句中 id 属性后面所对应的一定要与 AreaMapper 接口中对应的 selectByPrimaryKey 方法相一致，此属性是将对应的数据库操作语句与 Dao 中的接口类进行绑定。

对于每一个数据库操作语句而言，一般都遵循下面的格式：

```
1   <操作标签名称 id="Mapper接口中对应的方法" parameterType="数据类型" resultMap
    ="查询结果对应的结果集">
2       数据库操作语句
3   </操作标签名称>
```

其中，本章涉及的操作标签名称有 select、delete、insert 和 update。

2. "豹考通"数据库基本操作实现

接下来，我们针对上一节中列举的方法具体实现增删改查的基本操作。

（1）增加数据。首先打开 8.2 节中建立的 src/java/com.example.demo/controller 路径下的 TestController.java 文件。

由于要使用 AreaMapper 接口，因此在 Test 方法的外面首先初始化 AreaMapper 接口，并且通过注解@Autowired 进行标记（见程序清单 8-12）。

程序清单 8-12：Code0801\Demo1\src\main\java\com\example\demo\Controller\TestController.java

```
1   @Autowired
2   private AreaMapper areaMapper;
```

在 Test()方法外创建 GetTeacher()方法并设置其请求路径为/getTeacher,随后在 GetTeacher()方法中 new 一个实例变量 Area,通过 getter/setter 方法为 Area 变量赋值作为插入的数据,并且调用 AreaMapper 接口的 insert()方法,将数据插入数据库中(见程序清单 8-13)。

程序清单 8-13:Code0801\Demo1\src\main\java\com\example\demo\Controller\TestController.java

```
1    @RequestMapping("/getTeacher")
2    public void GetTeacher(){
3        Area area = new Area();
4        area.setAreaid((short) 101);
5        area.setAreaname("张三");
6        areaMapper.insert(area);
7    }
```

如图 8-28 所示,最后在启动类 DemoApplication 中添加注解@MapperScan("com.example.demo.dao"),告诉系统为我们扫描 com.example.demo.dao 包下的 Dao,并重启项目。

```
@SpringBootApplication
@MapperScan("com.example.demo.dao")
public class DemoApplication {

    public static void main(String[] args) { SpringApplication.run(DemoApplication.class, args); }

}
```

图 8-28　在启动类 DemoApplication 中添加注解

通过浏览器访问 http://localhost:8080/getTeacher,查看数据库便可得到插入的数据,如图 8-29 所示。

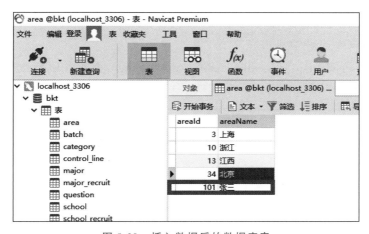

图 8-29　插入数据后的数据库表

(2) 查找数据。在 GetTeacher()方法中注释掉增加数据的代码,调用 AreaMapper 接口的 selectByPrimaryKey()方法,并重启项目,如程序清单 8-14 所示。

程序清单 8-14：Code0801\Demo1\src\main\java\com\example\demo\Controller\TestController.java

```
1    Area area = new Area();
2    area = areaMapper.selectByPrimaryKey((short)101);
3    System.out.println(area.getAreaid());
4    System.out.println(area.getAreaname());
```

通过浏览器访问 http://localhost:8080/getTeacher，如图 8-30 所示，通过程序控制台可以查看返回的数据。

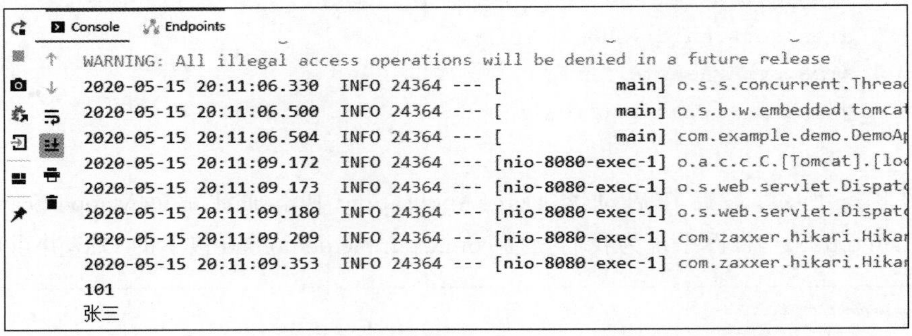

图 8-30　控制台显示返回的测试数据

（3）修改数据。在 GetTeacher() 方法中注释掉查找数据的代码，调用 AreaMapper 接口的 updateByPrimaryKey() 方法，并重启项目，如程序清单 8-15 所示。

程序清单 8-15：Code0801\Demo1\src\main\java\com\example\demo\Controller\TestController.java

```
1    Area area = new Area();
2    area.setAreaid((short) 101);
3    area.setAreaname("李四");
4    areaMapper.updateByPrimaryKey(area);
```

通过浏览器访问 http://localhost:8080/getTeacher，查看并刷新数据库，便可得到改变的数据，如图 8-31 所示。

图 8-31　数据库中的新增数据显示

（4）删除数据。在 GetTeacher() 方法中注释掉修改数据的代码，调用 AreaMapper 接口的 deleteByPrimaryKey() 方法，并重启项目，如程序清单 8-16 所示。

程序清单 8-16：Code0801\Demo1\src\main\java\com\example\demo\Controller\TestController.java

1   areaMapper.deleteByPrimaryKey((short)101);

通过浏览器访问 http://localhost:8080/getTeacher，查看并刷新数据库，便可得到改变的数据，如图 8-32 所示。

图 8-32　数据库中的删除数据显示

8.5　App 服务器业务逻辑层设计与实现

8.5.1　App 服务器业务逻辑层设计

业务逻辑层的关注点主要集中在业务规则的制订、业务流程的实现等与业务需求有关的系统设计。也就是说，它与系统所应对的领域逻辑有关，很多时候我们也将业务逻辑层称为领域层。

业务逻辑层在体系架构中的位置很关键，它处于数据访问层与表示层中间，在数据交换中起到了承上启下的作用。由于层是一种弱耦合结构，层与层之间的依赖是向下的，底层对于上层而言是"未知"的，因此改变上层的设计对于其调用的底层而言没有任何影响。如果在分层设计时遵循了面向接口设计的思想，那么这种向下的依赖也应该是一种弱依赖关系。因而在不改变接口定义的前提下，理想的分层式架构应该是一个支持可抽取、可替换的"抽屉"式架构。正因为如此，业务逻辑层的设计对于一个支持可扩展的架构尤为关键，因为它扮演了两个不同的角色。对于数据访问层而言，它是调用者；对于表示层而言，它却是被调用者。依赖与被依赖的关系都纠结在业务逻辑层上，如何实现依赖关系的解耦，则是除了实现业务逻辑之外留给设计师的任务。

针对我们的"豹考通"App 而言，一共有 4 个单独的页面，每一个页面对应自己单独的业务逻辑，本节将以图 8-33 所示的第一个页面"历年高考省控线查询页面"

图 8-33　"豹考通"历年高考省控线查询界面

作为案例进行讲解。

我们可以了解到本页面需要实现的业务逻辑为：根据输入的年份、生源地、批次、类别，查询与之对应的省控线数值。

根据以上业务逻辑，我们可以明确，需要调用的 Dao 为 ControlLineMapper。

8.5.2　App 服务器业务逻辑层实现

1."豹考通"基本功能类的实现

为了使业务逻辑层次更加清晰，我们需要在 src/java/com.example.demo 目录下新建一个包，命名为 service。然后新建一个 Java 类，命名为 ControlLineService 并给其添加注解@Service。

2."豹考通"功能类相互调用的实现

因为要调用 ControlLineMapper，所以我们需要首先初始化并添加@Autowired 注解。然后我们发现 ControlLineMapper 接口中只能通过主键查询省控线信息，因此需要在 ControlLineMapper 中手动添加 getControlLineByDetails 方法，根据传输的数据查询省控线，最后如图 8-34 所示。

```
public interface ControlLineMapper {
    int deleteByPrimaryKey(Integer controlid);

    int insert(ControlLine record);

    int insertSelective(ControlLine record);

    ControlLine selectByPrimaryKey(Integer controlid);

    int updateByPrimaryKeySelective(ControlLine record);

    int updateByPrimaryKey(ControlLine record);

    Short getControlLineByDetails(Short areaid , Short batchid , Short categoryid , Short controlyear);
}
```

图 8-34　在 ControlLineMapper 接口中添加 getControlLineByDetails 方法

此时，我们发现，如果要查询省控线，还需要知道省份的 id、批次的 id 和专业类别的 id，但对应的 Dao 中也是只能通过主键来查询，因此还需要在各个 Dao 中新建通过内容查询 id 的方法。

在 AreaMapper 中，我们需要添加如下方法：

```
1   Short getAreaIdByArea(String areaname);
```

在 BatchMapper 中，我们需要添加如下方法：

```
1   Short getBatchIdByBatch(String batchname);
```

在 CategoryMapper 中，我们需要添加如下方法：

```
1   Short getCategoryIdByCategory(String categoryname);
```

业务逻辑是根据输入的年份、生源地、批次、类别查询与之对应的省控线数值，因此我们在 ControlLineService 中创建一个方法 getControlLine()。

然后声明三个变量（areaid、batchid 和 categoryid），分别用来存储对应的 id 并分别执行各自的方法，具体代码如程序清单 8-17 所示。

程序清单 8-17：Code0801\Demo1\src\main\java\com\example\demo\service\ControlLineService.java

```
1    areaid = areaMapper.getAreaIdByArea("江西");
2    batchid = batchMapper.getBatchIdByBatch("一本");
3    categoryid = categoryMapper.getCategoryIdByCategory("理工");
```

这时我们需要打开各自的映射文件（AreaMapper.xml、BatchMapper.xml 和 CategoryMapper.xml），为其添加各自的查询语句。

以 AreaMapper 为例，我们需要首先为其添加一个新的结果集，用来确定查询的结果，代码如程序清单 8-18 所示。

程序清单 8-18：Code0801\Demo1\src\resource\mapper\AreaMapper.xml

```
1    <resultMap id="BaseResultMap1" type="java.lang.Short">
2        <id column="areaId" jdbcType="SMALLINT" property="areaid" />
3    </resultMap>
```

然后添加如程序清单 8-19 所示的查询语句。

程序清单 8-19：Code0801\Demo1\src\resource\mapper\AreaMapper.xml

```
1    <select id="getAreaIdByArea" parameterType="java.lang.String" resultMap="BaseResultMap1">
2        select
3        areaId
4        from area
5        where areaName = #{areaname,jdbcType=SMALLINT}
6    </select>
```

分别添加好相应的查询语句后，我们返回到 TestController 中，初始化 ControlLineService 并为其添加 @Autowired 注解，如图 8-35 所示。

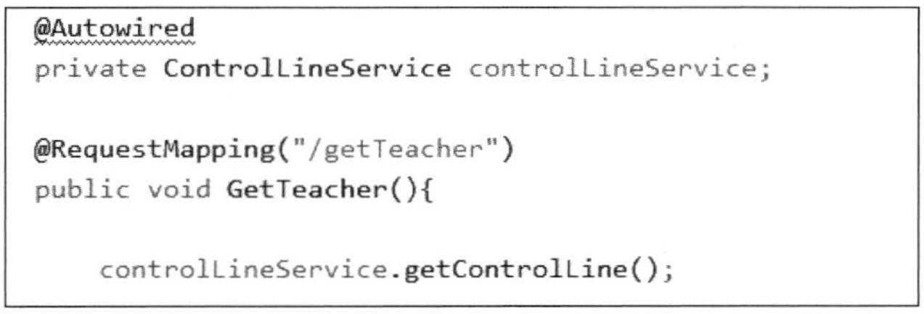

```
@Autowired
private ControlLineService controlLineService;

@RequestMapping("/getTeacher")
public void GetTeacher(){

    controlLineService.getControlLine();
```

图 8-35　ControlLineService 服务类

重启项目之后，看见控制台返回如图 8-36 结果则代表查询成功。

当成功获取三个 id 后，我们进入 ControlLineMapper.xml 中，首先为其添加结果集，代码如程序清单 8-20 所示。

```
2020-05-16 00:52:38.983  INFO 9204 --- [           main]
2020-05-16 00:52:39.175  INFO 9204 --- [           main]
2020-05-16 00:52:39.180  INFO 9204 --- [           main]
2020-05-16 00:52:41.184  INFO 9204 --- [nio-8080-exec-1]
2020-05-16 00:52:41.185  INFO 9204 --- [nio-8080-exec-1]
2020-05-16 00:52:41.191  INFO 9204 --- [nio-8080-exec-1]
2020-05-16 00:52:41.235  INFO 9204 --- [nio-8080-exec-1]
2020-05-16 00:52:41.397  INFO 9204 --- [nio-8080-exec-1]
13
1
2
```

图 8-36 ControlLineService 运行结果

程序清单 8-20：Code0801\Demo1\src\resource\mapper\ControlLineMapper.xml

```
1  <resultMap id="BaseResultMap1" type="java.lang.Short">
2      <resultcolumn="controlLine"jdbcType="SMALLINT" property="controlline" />
3  </resultMap>
```

然后添加如程序清单 8-21 所示的查询语句。

程序清单 8-21：Code0801\Demo1\src\resource\mapper\ControlLineMapper.xml

```
1  <select id="getControlLineByDetails" resultMap="BaseResultMap1">
2      select
3      controlLine
4      from control_line
5      where areaId = #{areaid,jdbcType=SMALLINT}
6      and batchId = #{batchid,jdbcType=SMALLINT}
7      and categoryId = #{categoryid,jdbcType=SMALLINT}
8      and controlYear = #{controlyear,jdbcType=SMALLINT}
9  </select>
```

最后，重启项目并在浏览器中刷新 http://localhost:8080/getTeacher 页面，控制台出现如图 8-37 所示结果则表示查询成功。

图 8-37 getControlLineByDetails 成功返回结果

8.6 App 服务器控制层设计与实现

8.6.1 App 服务器控制层设计

控制层的职能是负责读取视图表示层的数据，控制用户的输入并调用业务层的方法。

针对"豹考通"App 后台服务器而言，控制层的作用就是把用户从 App 界面上传递的值，通过 URL 的形式传递给后台，并获取，通过获取的信息，结合对应的 service 方法进行查找数据，从而获取数据库内的值。

根据上面的逻辑，我们可以知道，要设计 App 服务器控制层，首先需要创建一个控制器。然后通过控制器获取 URL 传递的参数，最后通过这些参数确定 Service 服务并返回相应参数对应的返回值。

整个过程如图 8-38 所示。

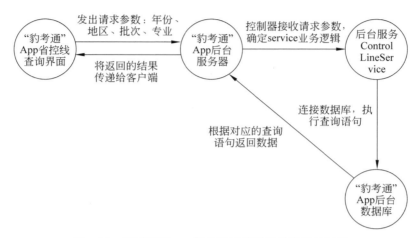

图 8-38 "豹考通"App 服务器控制层接收数据的过程

8.6.2 App 服务器控制层实现

根据上一节的设计，我们首先需要在 src/java/com.example.demo/controller 路径下新建一个类，命名为 ControlLineController，并且为该类添加注解@RestController，告诉 Spring Boot，ControlLineController 是一个控制器。

然后新建方法 getControlLine()，添加注解@RequestMapping(value="/getControlLine")；为 getControlLine()方法添加 4 个变量，用于接收前台提交的数据。

需要注意的是，在 Spring Boot 框架中，我们选择通过注解标注参数的方式实现后台对前台数据的接收。

具体操作方式是对参数进行注解，通过添加@RequestParam 注解的方式告诉 Spring Boot 后台，用于处理参数名对应数据，具体例子如程序清单 8-22 所示。

程序清单 8-22：Code0801\Demo1\src\main\java\com\example\demo\Controller\ControlLineController.java

```
1    /*http://localhost:8080/getControlLine?areaname=上海*/
2    @RequestParam(value = "areaname",defaultValue = "江西")
3    String areaname,
```

我们看到，在@RequestParam 注解中出现了两个参数。

value 参数的作用是将接收的参数名与实际参数进行绑定,内容必须和参数名称保持一致。defaultValue 参数的作用是确定 value 参数接收的默认值。

结合具体 URL 分析,我们可以看出,该控制器的 areaname 的值为上海;URL 中 areaname 为空时,则控制器的 areaname 的值为江西。

根据上面的例子,我们可以为 getControlLine()方法添加 4 个参数,分别是 areaname、batchname、categoryname 和 year,具体代码如程序清单 8-23 所示。

程序清单 8-23: Code0801\Demo1\src\main\java\com\example\demo\Controller\ControlLineController.java

```
1    public void getControlLine(
2    @RequestParam(value = "areaname",defaultValue = "江西")
     String areaname,
3    @RequestParam(value = "batchname",defaultValue = "一本")
     String batchname,
4    @RequestParam(value = "categoryname",defaultValue = "理工")
     String categoryname,
5    @RequestParam(value = "year",defaultValue = "2010")
     Short year)
```

最后调用 ControlLineService 的 getControlLine()方法,重启 Spring Boot 项目。在浏览器中输入下列 URL 路径 http://localhost:8080/getControlLine?areaname=上海&batchname=一本&categoryname=理工&year=2010,看到控制台返回图 8-39 所示数据则表示成功完成服务器的搭建。

图 8-39 ControlLineController 成功返回数据

8.7 本章小结

本章以 Spring Boot 基础项目的搭建、Maven 的配置、Mybatis 对数据库的管理为基础,通过讲解"豹考通"App 后台服务器省控线查询服务的具体设计与实现过程,让同学们通过真实的项目了解并学习如何使用"Spring Boot+Mybatis"框架搭建 App 服务器后台服务,同时做到前后台数据交互,实现对真实项目的参与,提高学生的学习兴趣与实际动手能力。

8.8 课后练习

1. 参照本章全部内容,自主独立完成"Spring Boot＋Mybatis"框架为基础"豹考通" App 后台服务器的搭建。
2. 结合 8.4 节独立设计并实现"豹考通"App 后台服务器剩余模块的数据持久层。
3. 结合 8.5 节独立设计并实现"豹考通"App 后台服务器剩余模块的业务逻辑层。
4. 结合 8.6 节独立设计并实现"豹考通"App 后台服务器剩余模块的控制层。

第 9 章

App 微信分享的实现

9.1 本章简介

在日常 App 开发中为了方便让客户帮助宣传本应用,一般提供分享的功能,其中微信分享最为常见。本章主要围绕实现微信分享功能,在"豹考通"教学版中添加代码进行实现。微信分享功能实现需要借助微信开放平台,在上面创建移动应用获取 AppID,完成填写基本信息、填写平台信息、提交成功 3 个阶段。其中难点在于填写平台基本中对 apk 安装包进行签名并获取签名信息才能注册。本章重点讲解 APK 包的签名过程、获取签名信息、"豹考通"教学版微信好友和朋友圈分享实现。

9.2 App 微信分享的操作流程

9.2.1 微信开放平台

分享到微信时需要用到微信开放平台,微信开放平台实际上是为第三方移动程序提供接口。用户可以在 App 上把看到的内容通过微信开放平台提供的接口发送给微信好友或分享至朋友圈,这样 App 上的内容就可以在微信平台获得更广泛的传播。Android 程序员要实现微信开发就需要在微信开放平台创建应用,如图 9-1 所示。

图 9-1 在微信开放平台创建应用

创建应用需要完成填写基本信息、填写平台信息、提交成功 3 个步骤,如图 9-2 所示。其中填写基本信息指的是填写要实现的应用的基本信息,包括移动项目名称、英文名(选填)、移动应用简介、英文简介(选填)、应用官网、移动应用图片、申请应用说明等信息。填写平台信息包括应用下载地址、应用签名、应用包名 3 个信息,如图 9-3 所示。以上信息中的一个难点是应用签名,我们将在 9.3 节中重点讲解。第三步是提交,等待微信开放平台管理员审核,如果出现问题,则需要修改后重新提交。

图 9-2　微信开放平台创建应用的步骤

图 9-3　填写平台信息

9.2.2　将 App 内容分享给微信好友

在 App 中想把一些截图、链接等分享给自己的微信好友时，通过点击"分享""发送给好友"按钮，就能把信息发送给微信好友。好友收到信息后，轻轻一点，就可以查看详情。另外，还可以使用微信 App 来查看内容（没有安装微信的用户将会被提示去下载安装），如图 9-4 所示。

图 9-4　将 App 内容分享给微信好友

9.2.3　将 App 内容分享到微信朋友圈

如果我们在 App 软件中想把一些截图、链接等精彩内容分享到微信朋友圈,则需要单击"分享""分享到朋友圈"按钮。完成授权后,内容就可以发送到微信的服务器。自己的好友在朋友圈中马上就能看到分享的内容。另外,还可以使用微信来查看分享的内容(没有安装微信的用户将会被提示去下载安装),如图 9-5 所示。

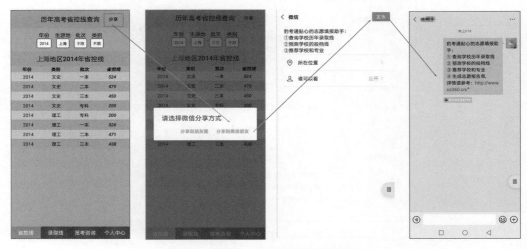

图 9-5　将 App 内容分享到微信朋友圈

9.3　Android 应用打包签名

Android 使用包名作为唯一标识,如果在同一台手机上安装两个包名相同的应用,则后安装的应用会覆盖前面的应用(签名相同的情况下)。

签名有两个主要作用:

- 确定发布者身份。由于应用开发者可以通过使用相同的包名来替换已经安装的程序,因此使用签名可以避免发生这种情况。
- 确保应用的完整性。签名会对应用包中的每个文件进行处理,从而确保程序包中的文件不会被替换。

在开发和调试阶段,Android Studio(AS)会自动生成调试证书进行签名。当应用发布时,必须使用合适的数字证书对应用进行签名。

作为谷歌在 2013 年为开发者提供的 IDE 环境工具 Android Studio,从几次更新之后 Android Studio 已经成为非常强大的 IDE 开发环境。谷歌也宣布 Android Studio 将取代 Eclipse。本节主要介绍如何使用 AS 生成签名文件、配置 gradle 让 App 自动签名以及如何生成 SHA1 和 MD5 值。

9.3.1 打包签名 apk 文件

(1) 在 AS 菜单栏中找到 Build 这一栏,单击选择 Generate Signed Bundle/APK 选项,如图 9-6 所示。

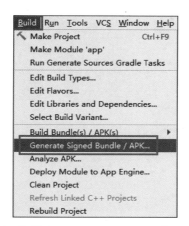

图 9-6 选择 Generate Signed Bundle/APK 选项

(2) 这里选择新建一个文件,如图 9-7 所示。

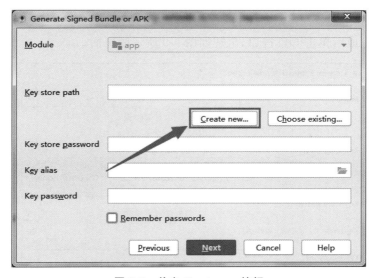

图 9-7 单击 Create new 按钮

- Create new 按钮表示新建一个签名文件。
- Choose existing 按钮表示选择一个已经存在的签名文件。

(3) 填写完整的签名信息(见图 9-8)。这里密码统一填写 android,别名填写 android 即可。其他信息根据实际情况和需求填写,并不是很重要。填写完信息在指定的路径下会生成 android.jks 文件,此文件就是我们所需要的签名文件。

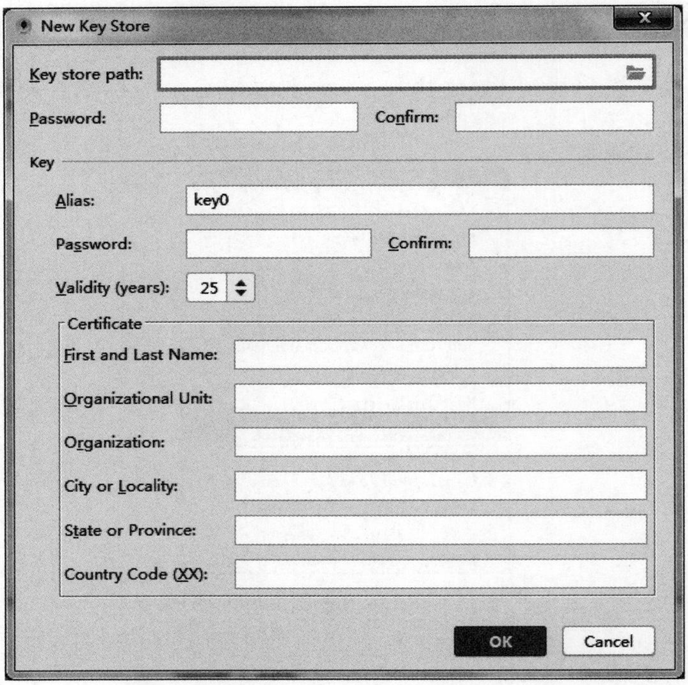

图 9-8 创建签名文件

（4）填写完成后单击 OK 按钮，出现如图 9-9 所示的页面。

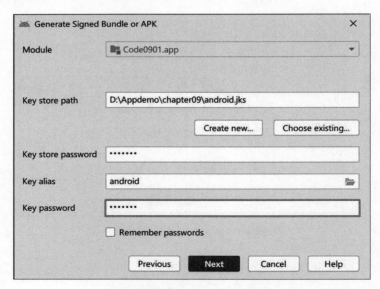

图 9-9 使用创建好的签名文件

（5）单击 Next 按钮，进入下一页面，选择 release 和 v2，然后单击 Finish 等候签名成功，如图 9-10 所示。

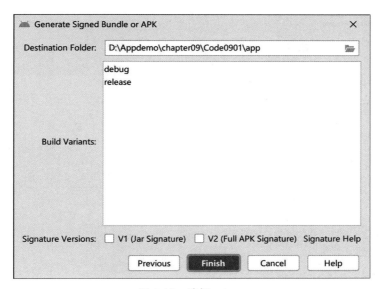

图 9-10　选择 release

- release 是发布版本用的签名文件。
- debug 是调试用的签名文件。

（6）等待 APK 文件导出完成，最终位置在项目的 app 文件夹的 release 文件夹下，如图 9-11 所示。

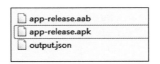

图 9-11　生成的 APK 打包文件

9.3.2　配置 gradle 让 App 自动签名

（1）在 AS 菜单栏中选择如图 9-12 所示图标。

图 9-12　选择适当图标

（2）选择 Modules 菜单栏→选中"＋"另添加标签。填写一个自定义的名称，同时根据新建签名文件时的信息将所有内容填写完整，如图 9-13 所示。

（3）填写完成后，单击 OK 按钮。当 Studio 编译完成后，会在 build.gradle 文件中自动生成配置信息，如图 9-14 所示。

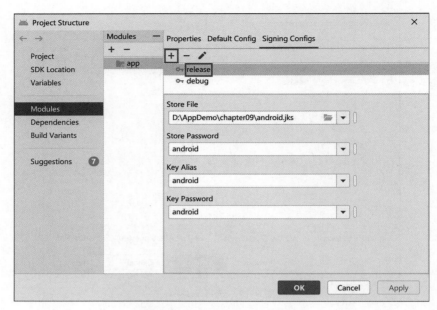

图 9-13　Project Structure 窗口

```
signingConfigs {
    release {
        storeFile file('D:\\AppDemo\\chapter09\\android.jks')
        storePassword 'android'
        keyAlias = 'android'
        keyPassword 'android'
    }
}
```

图 9-14　生成配置信息

9.4　Android 平台分享到微信的开发流程

9.4.1　申请 AppID

在 9.2 节已经介绍微信分享的操作流程，完成填写基本信息、填写平台信息、提交成功 3 个步骤。基本信息根据自己要实现的应用填写信息即可。填写微信平台信息里面需要写入开发应用的应用签名和应用包名，包名在 AndroidManifest.xml 中可以获取。在上一节中我们对应用已经实现签名，但是还未获得签名信息，下面介绍获取应用签名的两种方式。

（1）SHA1 和 MD5 值可以使用 DOS 窗口命令生成，在 AS 中我们可以直接使用 Terminal 工具。

① 在 AS 底部菜单栏中选择 Terminal 工具，如图 9-15 所示。

② 使用 DOS 命令将目录切换成 android.jks 文件目录（如图 9-16 所示），输入

第 9 章　App 微信分享的实现

图 9-15　Terminal 工具

keytool -list -v -keystore android.jks，按下回车键。输入 keystore 密码 android，结果如图 9-17 所示。其中 MD5 就是我们所需要的签名信息。注意，获取的 MD5 签名信息在微信开放平台填写时需要去掉其中的冒号。

图 9-16　切换到 android.jks 文件目录下

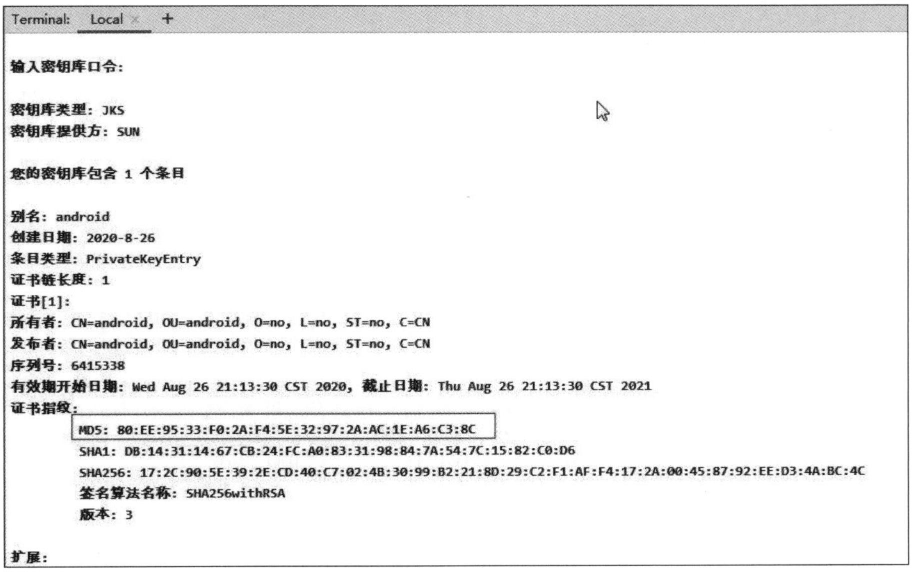

图 9-17　查看 MD5 签名信息

（2）利用签名生成工具获取。签名生成工具可以微信开放平台下载，其实就是一个 APK 应用文件，需要安装在手机上。自己编写的 Android 应用项目也要安装在同一个手机中。然后打开签名工具，输入包名，就可获取签名信息，如图 9-18 所示。

注意：

① 在使用签名生成工具时，AndroidManifest.xml 中的包名要和 build.gradle 中的 applicationId 的包名一致。

② 安装在手机上的 APK 是签名生成的 APK，而不是 debug.apk，否则获取的签名信息不一致。

获得签名信息后在微信开发平台就可以填写其他信息申请 AppId，平台提示在 7 日内完成审核，因此要静心等待。如果反馈结果中出现问题，可以修改后重新提交审核。

图 9-18　通过签名生成工具获取签名信息

9.4.2　搭建开发环境

首先从微信开放平台下载 SDK 开发工具包。过去使用 Eclipse 工具开发时只能下载 libammsdk.jar，导入项目才能开发；现在使用 Android Studio 工具既可以下载导入 libammsdk.jar，也可以在 build.gradle 中添加依赖下载工具包。

1. 方式一

（1）在"豹考通"项目中找到 libs 目录，将开发工具包中 libs 目录下的 libammsdk.jar 复制到该目录中，如图 9-19 所示。

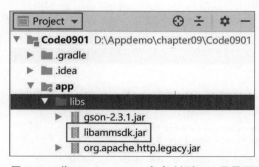

图 9-19　将 libammsdk.jar 包复制到 libs 目录下

（2）右击 libammsdk.jar，选择 Add As Library 选项（见图 9-20），将工具包导入项目（见图 9-21），同时在 build.gradle 中会出现添加依赖的情况，如图 9-22 所示。

2. 方式二

直接在 build.gradle 文件中添加 implementation 'com.tencent.mm.opensdk：wechat-

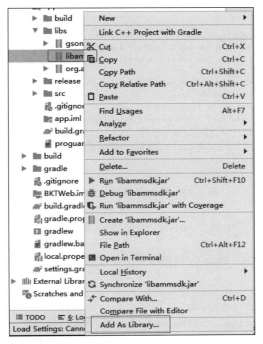

图 9-20　Add As Library

图 9-21　选择 app

```
dependencies {
    implementation fileTree(include: ['*.jar'], dir: 'libs')
    androidTestCompile('com.android.support.test.espresso:espresso-core:2.2.2', {
        exclude group: 'com.android.support', module: 'support-annotations'
    })
    implementation 'com.android.support:appcompat-v7:27.+'
    implementation 'com.android.support.constraint:constraint-layout:1.0.2'
    implementation files('libs/gson-2.3.1.jar')
    implementation files('libs/org.apache.http.legacy.jar')
    testImplementation 'junit:junit:4.12'
    implementation files('libs/libammsdk.jar')
}
```

图 9-22　将 libammsdk.jar 添加到依赖中

sdk-android-without-mta：+'，然后进行同步更新，即可下载微信开发工具包，如图 9-23 所示。

剩下的工作是进行微信分享代码编写，我们自己建立 WeChatShareUtil 工具类，用于

```
dependencies {
    implementation fileTree(include: ['*.jar'], dir: 'libs')
    androidTestCompile('com.android.support.test.espresso:espresso-core:2.2.2', {
        exclude group: 'com.android.support', module: 'support-annotations'
    })
    implementation 'com.android.support:appcompat-v7:27.+'
    implementation 'com.android.support.constraint:constraint-layout:1.0.2'
    implementation files('libs/gson-2.3.1.jar')
    implementation files('libs/org.apache.http.legacy.jar')
    testImplementation 'junit:junit:4.12'
    implementation 'com.tencent.mm.opensdk:wechat-sdk-android-without-mta:+'
}
```

图 9-23　添加微信开发 SDK 依赖

分享各种各样的内容（需要用到 AppID），如程序清单 9-1 所示。

程序清单 9-1：Code0901\app\src\main\java\cn\jxufe\iet\code\wxapi\WeChatShareUtil.java

```java
1   public class WeChatShareUtil {
2   //从官网申请的合法 AppID
3   public static final String APP_ID = "××××××××××";
4   private static final int TIMELINE_SUPPORTED_VERSION = 0x21020001;
5   //IWXAPI 是第三方 App 和微信通信的 openapi 接口
6   private IWXAPI api;
7   private Context context;
8   public static WeChatShareUtil weChatShareUtil;
9   public static WeChatShareUtil getInstance(Context context) {
10      if (weChatShareUtil == null) {
11          weChatShareUtil = new WeChatShareUtil();
12      }
13      if (weChatShareUtil.api != null) {
14          weChatShareUtil.api.unregisterApp();
15      }
16      weChatShareUtil.context = context;
17      weChatShareUtil.regToWx();
18  return weChatShareUtil;
19      }
20  //注册应用 id 到微信
21  private void regToWx() {
22  //通过 WXAPIFactory 工厂获取 IWXAPI 的实例
23      api = WXAPIFactory.createWXAPI(context, APP_ID, true);
24  //将应用的 AppID 注册到微信
25      api.registerApp(APP_ID);
26      }
27  /**
28    * 分享文字到朋友圈或者好友
```

```
29      * @param text     文本内容
30      * @param scene    分享方式:好友还是朋友圈
31      */
32     public boolean shareText(String text, int scene) {
33         //初始化一个WXTextObject对象,填写分享的文本对象
34         WXTextObject textObj = new WXTextObject();
35         textObj.text = text;
36         return share(textObj, text, scene);
37     }
38     /**
39      * 分享图片到朋友圈或者好友
40      * @param bmp      图片的Bitmap对象
41      * @param scene    分享方式:好友还是朋友圈
42      */
43     public boolean sharePic(Bitmap bmp, int scene) {
44         //初始化一个WXImageObject对象
45         WXImageObject imageObj = new WXImageObject(bmp);
46         //设置缩略图
47         Bitmap thumb = Bitmap.createScaledBitmap(bmp, 60, 60, true);
48         bmp.recycle();
49         return share(imageObj, thumb, scene);
50     }
51     /**
52      * 分享网页到朋友圈或好友,视频和音乐的分享和网页的分享大同小异,只是对象不同
53      * @param url            网页的URL
54      * @param title          显示分享网页的标题
55      * @param description    对网页的描述
56      * @param scene          分享方式:好友还是朋友圈
57      */
58     public boolean shareUrl(String url, String title, Bitmap thumb, String description, int scene) {
59         //初始化一个WXWebpageObject对象,填写URL
60         WXWebpageObject webPage = new WXWebpageObject();
61         webPage.webpageUrl = url;
62         return share(webPage, title, thumb, description, scene);
63     }
64     private boolean share(WXMediaMessage.IMediaObject mediaObject, Bitmap thumb, int scene) {
65         return share(mediaObject, null, thumb, null, scene);
66     }
67     private boolean share(WXMediaMessage.IMediaObject mediaObject, String description, int scene) {
68         return share(mediaObject, null, null, description, scene);
69     }
```

```java
70  private boolean share (WXMediaMessage.IMediaObject mediaObject, String
    title, Bitmap thumb, String description, int scene) {
71  //初始化一个 WXMediaMessage 对象,填写标题和描述
72          WXMediaMessage msg = new WXMediaMessage(mediaObject);
73      if (title != null) {
74          msg.title = title;
75          }
76      if (description != null) {
77          msg.description = description;
78          }
79      if (thumb != null) {
80          msg.thumbData = bmpToByteArray(thumb, true);
81          }
82  //构造一个 Req
83          SendMessageToWX.Req req = new SendMessageToWX.Req();
84          req.transaction = String.valueOf(System.currentTimeMillis());
85          req.message = msg;
86          req.scene = scene;
87  return api.sendReq(req);
88          }
89  //判断是否支持转发到朋友圈
90  //微信 4.2 以上支持,如果需要检查微信版本支持 API 的情况,可调用 IWXAPI 的
    //getWXAppSupportAPI 方法,0x21020001 及以上支持发送朋友圈
91  public boolean isSupportWX() {
92      int wxSdkVersion = api.getWXAppSupportAPI();
93      return wxSdkVersion >= TIMELINE_SUPPORTED_VERSION;
94          }
95  private byte[] bmpToByteArray(final Bitmap bmp, final boolean
    needRecycle) {
96          ByteArrayOutputStream output = new ByteArrayOutputStream();
97          bmp.compress(Bitmap.CompressFormat.PNG, 100, output);
98      if (needRecycle) {
99          bmp.recycle();
100         }
101 byte[] result = output.toByteArray();
102 try {
103         output.close();
104         } catch (Exception e) {
105         e.printStackTrace();
106         }
107 return result;
108         }
109 }
```

程序清单 9-2 所示为在项目下新建 wxapi 包,并在包下建立 WXEntryActivity.java

文件(微信的强制要求,用于接收微信的请求及返回值)。

程序清单 9-2:Code0901\app\src\main\java\cn\jxufe\iet\code\wxapi\WXEntryActivity.java

```java
1    public class WXEntryActivity extends Activity implements IWXAPIEventHandler {
2        @Override
3        protected void onCreate(Bundle savedInstanceState) {
4            super.onCreate(savedInstanceState);
5            IWXAPI api = WXAPIFactory.createWXAPI(this,
                 WeChatShareUtil.APP_ID, false);
6            api.handleIntent(getIntent(), this);
7            finish();
8        }
9        @Override
10       public void onReq(BaseReq baseReq) {
11       }
12       @Override
13       public void onResp(BaseResp baseResp) {
14           String result;
15           switch (baseResp.errCode) {
16               case BaseResp.ErrCode.ERR_OK:
17                   result = "分享成功";
18                   break;
19               case BaseResp.ErrCode.ERR_USER_CANCEL:
20                   result = null;
21                   break;
22               default:
23                   result = "分享失败";
24                   break;
25           }
26           if (result != null) {
27               Toast.makeText(this, baseResp.errCode + result,
                     Toast.LENGTH_SHORT).show();
28           }
29       }
30   }
```

在 manifest 中添加 WXEntryActivity 的相关 Activity 信息,如图 9-24 所示。

```
    </activity>
    <activity android:name="cn.jxufe.iet.code.client.ResultActivity"></activity>
    <activity android:name=".wxapi.WXEntryActivity"></activity>
</application>
```

图 9-24　WXEntryActivity 的相关 Activity 信息

9.4.3 实现微信分享功能

我们在省控线页面实现微信分享功能，首先需要在省控线布局导航栏中添加"分享"按钮，代码如图 9-25 所示。

```xml
<Button
    android:id="@+id/wechartshare"
    android:layout_alignParentRight="true"
    android:text="分享"
    android:background="#0000"
    android:layout_width="wrap_content"
    android:layout_height="wrap_content"/>
```

图 9-25 添加"分享"按钮

然后，如程序清单 9-3 所示，在 ControllineFragment.java 文件中实现按钮监听。在弹出对话框后，用户可以根据自己的需求，选择分享到微信朋友或朋友圈，运行效果如 9.2 节所示。虽然这里只介绍如何分享到微信朋友和朋友圈，但在 WeChatShareUtil 工具类中提供了分享图片和 URL 等的方法，开发者可以根据自己的开发需求进行调用。

程序清单 9-3：Code0901\app\src\main\java\cn\jxufe\iet\code\wxapi\WXEntryActivity.java

```java
1   weChatShareUtil = WeChatShareUtil.getInstance(getContext());
2   wechartShare.setOnClickListener(new View.OnClickListener() {
3       @Override
4       public void onClick(View v) {
5           //微信分享
6           AlertDialog.Builder builder=new AlertDialog.Builder(getContext());
7           builder.setTitle("请选择微信分享方式");
8
9           builder.setPositiveButton("分享到微信朋友", new DialogInterface.OnClickListener() {
10              @Override
11              public void onClick(DialogInterface dialog, int which) {
12                  //分享给微信朋友
13                  boolean result = true;
14                  sessionText = "豹考通贴心的志愿填报助手:\n①查询学校历年录取线\n②预测学校的投档线\n③推荐学校和专业\n④生成志愿报告单,\n 详情请参考:http://www.xs360.cn/\"";
15                  result = weChatShareUtil.shareText(sessionText, SendMessageToWX.Req.WXSceneSession);
16                  if (!result) {
17                      Toast.makeText(getContext(), "没有检测到微信", Toast.LENGTH_SHORT).show();
18                  }
19              }
```

```
20              });
21              builder.setNegativeButton("分享到朋友圈", new 
        DialogInterface.OnClickListener() {
22                  @Override
23                  public void onClick(DialogInterface dialog, int which) {
24                      //分享到朋友圈
25                      boolean result = true;
26                      if (weChatShareUtil.isSupportWX()) {
27                          String text = "豹考通贴心的志愿填报助手：\n①查询学校历年录
        取线\n②预测学校的投档线\n③推荐学校和专业\n④生成志愿报告单,\n详情请参考：
        http://www.xs360.cn/\"";
28                          result = weChatShareUtil.shareText(text,
        com.tencent.mm.sdk.modelmsg.SendMessageToWX.Req.WXSceneTimeline);
29                      } else {
30                          Toast.makeText(getContext(), "手机上微信版本不支持分
        享到朋友圈", Toast.LENGTH_SHORT).show();
31                      }
32                  }
33              });
34              builder.show();
35          }
36      });
```

9.5 本章小结

本章介绍了"豹考通"微信分享功能的实现。在实现微信分享功能之前，需要做一些准备工作。首先在微信开放平台创建应用，其中包括填写基本信息、填写平台信息、提交成功 3 个步骤。然后需要创建属于自己的签名文件、对本项目 APK 进行签名并获取签名信息。提交申请直到申请到 AppID 为止。

申请到 AppID 后，对项目进行开发。首先搭建开发环境，在微信开放平台下载 SDK 资源包添加到项目中，本章提供了两种方式。为了方便微信分享功能的实现，编写工具类 WeChatShareUtil.java 和 WXEntryActivity.java。在省控线查询模块中添加"分享"按钮，实现分享到朋友圈和微信朋友功能。

9.6 课后练习

1. 在微信开发平台注册 Android 微信分享开发的 AppID。
2. 在微信平台上分别下载 Android 微信分享开发工具，搭建 Android 微信分享开发环境。
3. 思考"豹考通"微信分享设计的思路，提出自己可以改进的见解。
4. 从微信开发平台下载 Android 平台参考手册或微信 SDK Sample Demo 源码查看相关内容。

图书资源支持

感谢您一直以来对清华版图书的支持和爱护。为了配合本书的使用,本书提供配套的资源,有需求的读者请扫描下方的"书圈"微信公众号二维码,在图书专区下载,也可以拨打电话或发送电子邮件咨询。

如果您在使用本书的过程中遇到了什么问题,或者有相关图书出版计划,也请您发邮件告诉我们,以便我们更好地为您服务。

我们的联系方式:

地　　址:北京市海淀区双清路学研大厦 A 座 714

邮　　编:100084

电　　话:010-83470236　010-83470237

客服邮箱:2301891038@qq.com

QQ:2301891038(请写明您的单位和姓名)

资源下载: 关注公众号"书圈"下载配套资源。

资源下载、样书申请

书圈

图书案例

清华计算机学堂

观看课程直播